GEMA: Birthplace of German Radar and Sonar

GEMA: Birthplace of German Radar and Sonar

Harry von Kroge
Hamburg, Germany

Translated from the German and edited by Louis Brown

CRC Press
Taylor & Francis Group
Boca Raton London New York

CRC Press is an imprint of the
Taylor & Francis Group, an **informa** business

CRC Press
Taylor & Francis Group
6000 Broken Sound Parkway NW, Suite 300
Boca Raton, FL 33487-2742

First issued in paperback 2019

No claim to original U.S. Government works

ISBN-13: 978-0-367-45553-8 (pbk)
ISBN-13: 978-0-7503-0732-1 (hbk)

Visit the Taylor & Francis Web site at
http://www.taylorandfrancis.com

and the CRC Press Web site at
http://www.crcpress.com

CONTENTS

Contents

FOREWORD

The location of targets by means of sound and radio waves represents an important part of the many technical advances of this century. I had the good fortune to experience this work from the beginning with my friend, Hans-Karl Freiherr von Willisen, and to have contributed something to its development. It was our privilege to have enthusiastic scientific and technical co-workers at our side with whom we quickly converted Dr Rudolf Kühnhold's ideas into practice. In the course of this work we learned that Christian Hülsmeyer had successfully demonstrated thirty years earlier the location of ships with the primitive radio methods of the time. The importance of his invention was not realized at the time, and he found no employment for his 'Telemobiloskop'.

A decisive contribution to our activities at GEMA was made by our scientific co-workers Dr Theodor Schultes, leader of the high-frequency laboratory, and Dr Walter Brandt, leader of the low-frequency laboratory. They were model team-chiefs from whom fundamental developments in radar and sonar originated.

I am grateful and pleased that Harry von Kroge has finally, after long years of research, been able to tell this detailed and significant story.

I hope this book with its report of an interesting part of German technical history achieves a well-deserved success.

Paul-Günther Erbslöh
Berchtesgaden 1998

TRANSLATOR'S PREFACE

The story of radar's invention and development in the years preceding and during World War II is a confused one, with serious gaps in a record confounded by error and tinged with mythology. This unfortunate state of affairs is the direct consequence of the severe secrecy that was imposed on this strange new vocation. For many years the people writing the history of radar had been participants in its formation and maturation, a circumstance that gave weight to their presentations but that invariably skewed their articles and books toward narrow points of view. Given that even the descriptions of the Allied efforts are covered in a highly irregular way, it is hardly surprising that those of the Germans and Japanese are but poorly covered, and are certainly not well known to English-language scholars. Many of their records were destroyed during the war, and the industrial research communities that built this equipment were disrupted and scattered. Their immediate postwar world left them with problems more pressing than writing history.

Former Telefunken people have given their company, famous for the Würzburg gun-laying and Lichtenstein airborne radars, some coverage, but the origins of the equally famous Freya air-warning and Seetakt ocean-surveillance radars are confused, even in Germany. It is generally reported, although not correctly, that their manufacturer, GEMA, an acronym for Gesellschaft für elektroakustische und mechanische Apparate, was only a cover for ownership by the German Navy. That they built the first functioning radar is appreciated almost nowhere; that their equipment during the first years of the war was the best of its kind will be disputed hotly by many, but the evidence justifies such a statement. But GEMA personnel wrote no histories.

Sonar does not rank in importance with radar, but histories of it are entirely from the British and American viewpoints. GEMA also designed and manufactured Germany's sonar and the development of the two devices is remarkably interrelated.

The reader will find two valuable topics discussed here: (1) a technical description of the evolution of GEMA's radars and sonars, and (2) a detailed history of an industrial corporation in wartime Germany. The former is instructive in comparing the development of similar equipment among the Allies. The second presents remarkable and rare material about how industry was carried out under such singular operating conditions.

The author has devoted decades to seeking out persons important to GEMA's history and in so doing uncovered a wealth of widely scattered documentary material, generally in the hands of individuals, not in formal archives. His work became known to some of us outside Germany a few years ago in manuscript form and was immediately recognized for its importance. The late Dr John H Bryant, not being fluent in German but seeing the great value of the material, had the text translated by Dr David Learned, MD, who spoke it sentence by sentence into a tape recorder. The results were transcribed and edited by Bryant. This translation, though useful to Bryant, was not suitable for publication and has not been used or referred to in providing this version.

The reader will find that we have not been consistent in dealing with names of organizations and apparatus. A translated form is given for some, others are left in German when the word or phrase has or develops currency during the reading and where a substitute English word proved to be awkward. The Subject Index can be used to relate the two.

One cannot present the history of radar without touching on technical matters to some extent, and the level at which to tell the story is a difficult problem for the author. Radar history has lessons valuable for electrical engineers, physicists and military personnel as well as historians concerned with technology, industry, armaments and strategy. Add to them laymen of wide interests, and the author will be puzzled as to what to include. The historian will be confused and irritated by extensive technical descriptions; the engineer will react similarly to their omission. Some technical matters are aimed at all readers, but other parts that are judged of little interest except to engineers and physicists are marked in **boldface** and can be skipped without losing the thread.

The translator and author wish to thank Mr F A Kingsley and Mr John Street for their careful reading of the text. Their comments resulted in numerous improvements.

Louis Brown
Washington, 23 September 1998

INTRODUCTION

A number of circumstances caused me to probe into the history of a corporation that had considerable importance for radar technology in Germany from its beginning until 1945. I had an interest in radio even in my school days. During and immediately after the war I lived with my grandparents on a farm in the so-called Kehdinger region between Stade and the mouth of the Oste on the lower Elbe. There were in that region numerous air warning and fighter control stations constructed for defence and that I could reach either on foot or on my bicycle. The stations were equipped during the course of the war with ever more complicated communication and radar apparatus, which I found particularly interesting.

In retrospect it is scarcely understandable how I, especially after 1944, could observe this equipment unmolested for hours, sometimes for days, equipment covered by the strictest secrecy. I watched not only the installation, routine operation, maintenance and crew training but also saw operation in response to air attack. I can perhaps explain this by noting that in 1944 the sharp young soldiers of the Luftwaffe air-warning service had been removed for more heroic duties at the front and replaced by men unsuitable for service there and by women of the Luftwaffe and the Reichsarbeitsdienst (National Labour Service), whose friendship I gained through gifts of produce from my grandfather's farm, which kept the friendly group in good humour. I even visited Flakhelfer (teenage soldiers) at an antiaircraft gun position in the uniform of these young people in order not to stand out.

Many experiences made a strong impression on me, especially in connection with modification and expansion when a lot of equipment could be seen. I gained a rough understanding of the various pieces and their function by watching and listening to the maintenance troops and company technicians. In February 1945 I was able to see the transmitter of a panoramic radar equipped with a spark modulator installed and tested. My experience with spark coils in physics class made this appear antiquated. I remembered that, besides the names of the well known producers, there appeared the name GEMA and that the equipment of this company carried the marking 'bya' rather than a company logo.

1

GEMA: Birthplace of German Radar and Sonar

I experienced the end of the war on 1 May 1945 when the town of Stade was given up without a fight and British troops occupied the Kehdinger land at the mouth of the Elbe. As it had been ordered earlier that Hamburg was to be defended, many German soldiers on the west bank of the river crossed over quickly. Inasmuch as there was no orderly retreat, a number of vehicles were abandoned, some being left on our property near the Elbe. Next to partially destroyed trucks stood two vehicles that were loaded with equipment that I recognized from my visits to the radar positions. With the help of a Polish prisoner whom I knew well I was able to salvage some of it and hide it from the British troops. It must have been a lasting impression formed earlier that caused me to pick apparatus bearing the marking 'bya' for hiding. The hiding place was so good that it was never discovered by the changing occupation troops in our vicinity. The hidden sets took up plenty of room and some weighed hundreds of pounds. Only my youthful unconcern allowed me to sleep well. I would not have been the only one to suffer, if my secret had come to light.

In the remains of a workshop in a destroyed part of the Hamburg Naval Arsenal and at a damaged radar position I found, before the British, the elements from which I gained my first knowledge. I discovered the meaning of the acronym 'GEMA' to be 'Gesellschaft für elektroakustische und mechanische Apparate m.b.H.', that they were located in Berlin, and I was able to identify part of my carefully listed equipment. I cautiously put my questions to soldiers of a technical company of a Luftwaffe research institute, who had been interned by the British in our vicinity in the summer of 1945. Because one could not continue lessons in secondary school in town until the fall, my mother allowed me to take private lessons from suitable members of this company. I found sufficient trust in one of these teachers to show him part of my secret hoard. At night I moved by myself the components of a GEMA N-Gerät into a remote room. In meetings that were supposed to cover my school studies my 'tutor' taught me the secrets of this equipment, a GEMA receiver and indicator. For the first time I received exact information about a part of my treasure.

(I insert here the reason why a company name was encoded in two or three letters: for reasons of secrecy the name of the producer could only be given in coded form. The Heereswaffenamt (Army Ordnance Department) in Berlin from 1940 until 1945 produced lists of manufacturers of weapons, munitions and equipment as secret documents, from which the name and address of the producer could be found.)

After capitulation the situation under British occupation loosened relatively quickly. Our vicinity was completely freed by the end of 1945, and the last German military personnel were released. I ventured slowly to bring my equipment out of hiding, and things again went to my advantage.

At the beginning of 1946 my grandfather employed a farmhand, who had worked during the war in radar service. He was by training a pastry cook and his radar knowledge came only from Luftwaffe drill. Although not an expert, he enhanced my knowledge as best he could and together we studied my equipment.

The solid construction always pleased me, even when the consequent weight plagued me. We learned that I possessed the electronic basis of a GEMA Freya radar. There were two or three units for some items, and in boxes were packed diverse receivers, transmitters and cathode-ray tubes.

Under the leadership of my physics teacher, nicknamed 'Pikkolo', and with the help of friends I was able to put the receiver and indicator into operation, but of course, there was nothing to receive. By inserting another receiver and making a small alteration, I was able nevertheless to receive the signals from Hamburg's first ultra-short-wave transmitter in 1949. This seemed legal to me, and I demonstrated it to everyone, both acoustically and visually. I would have liked to try out the transmitter but refrained, because it was forbidden and I was afraid of the high anode voltages, which was just as well. I was able to control my passion for unqualified tinkering and lovingly cared for the equipment, for which I had my reward years later.

In 1952 I became active in the Institute for Ionospheric Research in the Max Planck Institute as part of my engineering studies. There Professor Walter Dieminger, scientist and leader of the institute, gave my professional career its final stamp. In his laboratory I found what I wanted: radar technology in its purest form with equipment to use. My first work brought me into contact with GEMA equipment and the peculiarity that had shocked me earlier because I had held it to be antiquated: the spark gap.

High-power pulse transmitters were needed in order to measure the night-time propagation paths of the former Northwest German Broadcast network. For this purpose transmitters from the GEMA radar Freya were rebuilt with 500 kW spark-gap modulators, a technique proposed during the war by Professor Erwin Marx for high-power GEMA modulators, the installation of which I had watched with amusement at one of the radar stations I had visited.

I gained at this laboratory a unique introduction to the theoretical and practical techniques of transmission and reception, as well as of timing or ranging for ionosphere sounding. Through discussions with the accomplished people there, and by perusing the abundant documentary material, I was able to complete my knowledge of GEMA equipment. After altering my own equipment to the 2 m band, on a frequency for which I had earned an amateur license, I succeeded in determining the range and direction of aircraft with my GEMA equipment, using a home-made Yagi antenna. This kind of operation was, of course, not allowed with my license, so I tried it out only briefly. After that I used this modified set for longer periods to sound the ionosphere with single short-wave pulses.

Further experiments with my GEMA set had to be abandoned when I began work in Hamburg. Once I had moved part of my heavy equipment, but never again. I had to dispose of such gear but could not bring myself to part with the GEMA range-measuring delay unit marked 'Messkette' (measuring chain) or 'UK' until I had calibrated it with modern equipment. One such circuit that came from the earliest years of the company allowed me to convince myself of the technical mastery the company had realized.

Both professionally and personally I have examined many pieces of radar apparatus manufactured by competent firms in the period up to 1945 and do not want to make comparisons, as all were marked by German precision and workmanship, but GEMA products were known from the beginning for their enduring and robust construction.

The decision to pursue the history of GEMA came to me in the 1960s through a discussion with Dr Hans Rindfleisch, at the time Director of North-West-German Radio Company in Hamburg. I received from him the first concrete information about the connection between the former Naval Nachrichtenmittel-Versuchsanstalt (NVA, communications research laboratory) in which Dr Rindfleisch had been active, and the firm Tonographie, from which GEMA had grown, founded by the two owners of Tonographie, Paul-Günther Erbslöh and Hans-Karl Freiherr von Willisen. I learned from him that GEMA had been liquidated in 1945 and that there had been an associated company Tonographie, which von Willisen then headed in Wuppertal.

I am still moved today by the moment when Dr Rindfleisch mentioned Tonographie, a firm in which I then had professional connections, and soon my discussions were extended to none other than von Willisen himself. Subsequently I had many conversations with him about GEMA. He had attempted to put together a history of the company, but his premature death in 1966 prevented the completion of this task, and this unexpected death put an altogether too early end to our extensive conversations. He and Dr Rindfleisch have earned my thanks.

It has been tedious to ascertain the exact chronology of GEMA's development. The limits of the collaboration with the various naval units, especially with Dr Kühnhold in Kiel, were broad and ill defined. It was my special desire to recognize in a proper manner this collaboration as well as that with other companies and scientific institutes.

At home and abroad there have been many publications about radar. In general they represent GEMA as a cover for the Navy's production of radio-location equipment. This statement comes superficially from the early collaboration with the NVA, but it is not true. This situation gave me an additional reason to probe deeper into the firm's history.

From the many discussions that I had and from the many authentic documents that I located during my studies, I obtained the information needed for a history that would closely approximate the truth. Besides von Willisen, other GEMA personnel have provided much support through their interviews and written memoirs and by furnishing documents and pictures. I especially want to thank Mr Erbslöh. I also want to thank Dr Walter Brandt, who made himself so valuable to the company. There were also plenty of special reports from Mr Röhrig and Mr Henke for which I am especially grateful. Mrs Rhein, von Willisen's secretary for many years, was able to locate important documents from the papers of her former employer. For her help I am much obliged. Until Germany's reunification it was impossible to conduct research at the company's previous location in East Berlin, but a very agreeable working relationship developed after the archives of east and west were joined. Dr Czihak and his co-workers deserve my thanks for the splendid way they helped in my search for old papers.

After gathering the data and facts about GEMA, it was not easy to avoid a comparison with German and international literature, but that is not my intention. Indeed I shall be pleased if my presentation allows the reader to make this comparison for himself. To this purpose I have worked out the continuity and timing more accurately than might seem necessary.

In order to prevent the size of this book from exceeding acceptable limits, only the crucial phases of the company's evolution and of equipment development are included. The diversity of the apparatus developed and manufactured by GEMA does not allow one to describe it all. It is my intention to make this good for the interested reader with a later supplement.

CHAPTER 1

IN THE BEGINNING WAS AN IDEA

After his graduation from Göttingen in physics Dr Rudolf Kühnhold accepted a position in 1928 as scientist at the Navy's Nachrichtenmittel-Versuchsanstalt (NVA). This institution had been organized at Kiel in 1923 for research and development of naval communications. An auxiliary station was found at Pelzerhaken near Neustadt on the Lübecker Bucht (Lübeck Bay).

Underwater sound was, from the start, a special assignment for NVA. In addition to improving the performance of the existing acoustic telegraph and developing directional capability for underwater noise receivers there was an urgent need for a method of directing naval fire at surface or underwater targets using sound for direction and ranging. This task had been given to it by the Torpedoversuchsanstalt (Torpedo Research Laboratory, TVA), which needed such a device for the precise aiming of its missiles.

This was a challenge for the young scientist, who became the leader of the acoustic group. He could draw on very little previous work, and much basic research would be required for such a project. First the propagation of sound in water had to be understood in order to develop the theoretical framework on which the design depended for the complex acoustic, electric and mechanical systems that would transmit and receive directed sound waves.

The NVA had limited resources for conducting the necessary experiments and building such new equipment, and Dr Kühnhold was dependent on qualified firms for their collaboration. The first fundamental measurements used a fathometer from the Atlas Company, which was already in use by the Navy. At the same time they acquired a 30 kHz sound generator in an experimental stage with a transducer from the French company SCAM, according to Langevin-Florison. This work proceeded slowly because only a few boats could be made available and the weather often interrupted. Time was also lost when it was realized that the lower salt content of the Baltic relative to that of the North Sea had important effects on their data. Nevertheless, Kühnhold's group succeeded by 1933 in gaining a substantial understanding of the propagation, diffraction and echoing of sound in water and a knowledge of the natural noises of the ocean.

These experiments showed that the demands of directing the fire of naval ordnance could not be met with existing sound equipment. The TVA insisted on an accuracy of 1° at a range of 7.5 km. To achieve this required a completely new kind of apparatus. Kühnhold did see the realization of his ingenious idea for determining the lateral deviation of a target by using two separated microphones connected to amplifiers that showed respectively the sum and difference of the signals. In 1931 he patented this sum–difference method, which was employed not only for sound location but later also for radar.

Beginning in 1933 Kühnhold, who had become Scientific Director of NVA by then, had decided to try radio waves in place of sound for target location. The Pintsch Company in Berlin was developing, with the assistance of Professor Karl Kohl, a 13.5 cm transmitter and receiver using Barkhausen positive-grid tubes, the transmitter yielding 0.1 W continuous power. NVA ordered this equipment from Pintsch and had the transmitter and receiver mounted to parabolic-mirror antennas of six-wavelength diameter. In early fall 1933 the first experiments were carried out between the NVA building and the naval arsenal, located about a mile away. The transmitter stood on the NVA building, and its signal was easily received across the harbour at the arsenal. Reflections from either the building or ships were not observed, no matter how the antennas were rotated relative to one another.

It was a great disappointment when there was no evidence of reflections when the antennas were pointed directly at large ships. Kühnhold blamed these negative results on the low power of the Pintsch transmitter, but an immediate increase in the power was impossible, owing to inherent limitations of the Barkhausen tube. Professor Kohl estimated that the time needed to obtain high-power tubes for centimetre waves was a decade, but Kühnhold did not allow himself to be discouraged by these unsuccessful reflection experiments, and he placed an order with Pintsch to improve the power of their transmitter and the sensitivity of their receiver.

In addition to the experiments in collaboration with Pintsch, Kühnhold discussed the problem with established electronics companies, in particular with Telefunken. Because no one realized the depth of his ideas and thoughts, and because reflection experiments with centimetre waves were considered out of the question at the low power levels then available, he found no support.

CHAPTER 2

THE ORIGINATORS

Shortly after the end of World War I two Potsdam students discovered a common interest in wireless telegraphy and telephony. Paul-Günther Erbslöh, born in 1905 in Düsseldorf, came to Potsdam in 1917 through the transfer of his father, a high-level state official. Hans-Karl Freiherr von Willisen, born in 1906 in Berlin-Charlottenburg, grew up in a Prussian officer's family and studied with Erbslöh at the Realgymnasium in Potsdam. After the war von Willisen, by way of an intermediary friend of his parents, came into possession of a Type E 170 intermediate-receiver of the German Telephone Company (DE-TE-WE). During the war this receiver had been part of an Army wireless station and allowed earphone reception in the 600 to 150 m band (500–2000 kHz). Not only that, young von Willisen received a box that was richly filled with AEG and TKD triodes and many other radio and military components, which impelled him and his friend Erbslöh toward intensive electronic activity.

Von Willisen connected his receiver to various antennas and listened to what was on the air in the bands available, and Erbslöh, after various more or less successful experiments, built a similar receiver. They used the army triodes to build an amplifier that markedly raised the sensitivity of the receivers. After many failures and much effort the two mastered the techniques of feedback, which allowed them to build a regenerative receiver that left the sensitivity of their old sets in the shade and opened enormously the range of wireless signals they could hear. Improvements that could not be made from the provisions of their box of spares could be made through the purchase of components at nearby electrical stores, there having been many stores in Berlin that sold components from surplus military radio apparatus.

Von Willisen had learned Morse code, so it was inevitable that they should read commercial wireless traffic, both domestic and international, which led them quickly to discovering the existence of the stations at Nauen and Norddeich. After a number of unsuccessful bicycle trips von Willisen finally gained access to the transmitter personnel at Nauen and used them to quench his thirst for knowledge about the generation and radiation of radio waves. He had already learned the basis of continuous-wave transmitter operation from experiments with feedback.

The two heard the broadcasts of the Königswusterhausen station from the very beginning, which they used to improve the quality of the audio reception of their sets. They were, however, unable to establish personal contact with the operators, as they had at Nauen.

The first official German radio broadcasts came at the end of October 1923 from the studio of the phonograph record company VOX and were heard by the fifth-form boys, Erbslöh and von Willisen. Through their practice as amateurs and from their collecting and studying of foreign and domestic publications and bulletins, they gained an ever-increasing knowledge of the field of electronics as well as a good understanding of the basic physical principles. None of this was beneficial to their school work.

Von Willisen became a wizard at quickly finding people competent to answer their questions about problems that arose and extracting answers from them. In addition to transmitter and receiver technology he became interested in sound reproduction and built an amplifier and several microphones that worked rather well. A few days after the first broadcast von Willisen was able to establish contact with the technical leadership of VOX-House, which consisted of Director Scheffer and Engineer Heckmann. He and Erbslöh were able to participate in the trials of the recording apparatus of this station. After that it proved fairly easy to establish contact with the leader of the Königswusterhausen station, Erich Schwarzkopf. With Erbslöh he designed and built, among other things, a microphone that was made of three microphone capsules of the carbon button type, which were in common use in telephone sets. Its frequency response was corrected through tedious threading and stuffing of the region between the covering grid and the carbon diaphragm with wool. They embedded the three thus-improved capsules in felt surrounded by a wooden ring on a dowel as a handle. The microphone resembled visually the signal upheld by a railway platform master to indicate 'proceed' and was named just that: 'Abfahrtskelle'. The quality notably exceeded that of previous microphones.

Erbslöh and von Willisen found a friend in Erich Schwarzkopf, who was interested in an electrical pick-up from phonograph records, and engaged the two with his ideas, and in a remarkably short time the two had developed one. They cleverly mounted a needle onto the diaphragm of a telephone earphone, and the electrical pick-up was born. The converting of phonographs from acoustical to electronic soon brought them a nice income, and they often sold amplifiers and loudspeakers too.

Erbslöh and von Willisen then easily reversed the process. With a sharp sapphire mounted on the telephone earphone diaphragm they cut records on blank wax discs by applying the output voltage of an amplifier to the device. Since the time of Edison, Berliner and Pathe, discs had been recorded by purely acoustic methods, and corrections to the recorded frequencies were not possible. Recording an orchestra was done with a huge funnel that channelled the sound waves to the recording head, which was limited in its acoustic range. The two young engineers tried out various combinations of microphones and amplifiers with recording technicians from Deutsche Grammophon and Homophon. They impressed these two firms with their method of using multiple microphones and mixing the signals in a manner that later became standard with sound-recording engineers. With this technique they could combine various microphone channels and reproduce the sound of an orchestra substantially better than with the old acoustic technique.

Unconstrained by other commitments (other than school), Erbslöh and von Willisen pushed the development of their own recording heads for wax discs. They were uninterested in the turntable mechanisms, as these wind-up devices were well made by two companies, Mauritz and Neumann. The goal they had set themselves was to preserve music on a phonograph record and deliver it to the listener with the best possible fidelity.

They divided the work according to their individual abilities: Erbslöh worked on the recording head and von Willisen on a microphone amplifier with a power of 30–40 W to drive the head. The amplifier was also able to compensate for the frequency response of the recording head.

Today this kind of problem is easily solved through the use of a large variety of suitable measuring instruments not then in existence, and many fabrications and trials were required to attain their goal. The two were interrupted and delayed in this work by the need to graduate from school, which required study and a few semesters spent in Munich and Berlin. There was also the problem of securing sufficient financial support, obtained primarily through the sale of equipment that they made themselves. In a small workshop in Berlin-Charlottenburg Erbslöh produced many versions of his recording heads, a tedious way to find the right construction to fulfil their hopes. He worked through many a night with von Willisen, employing every bit of information that could be found. In addition to the lively exchanges with their friends at the Neumann Company, especially with Georg Neumann himself, they opened up fundamental discussions with various scientists at the Heinrich Hertz Institute for research into electromagnetic waves, under Professor K W Wagner. Von Willisen had built up this connection, and it was later to prove of great value. Professor Friederich Trautwein, who developed the so-called 'Trautonium', the predecessor of later instruments for electronic music, was often called on for advice.

The practical results of their newly developed sound systems gained Erbslöh and von Willisen connections to the Musikhochschule in Berlin-Charlottenburg, whose rector, Professor Schünemann, was open minded concerning new techniques. Herbert Grenzebach had a teaching position there and was greatly pleased with the efforts of the two inventors. After much long, laborious work they succeeded with their microphone and recording machine in reproducing music of high quality.

The two also worked for the sound studio Tonographie, thereby establishing contact with Mr Küchenmeister from the company of the same name, to which Ultraphon in Berlin also belonged, which made phonograph records that Küchenmeister sold. He learned about Erbslöh and von Willisen from the Charlottenburg music school and from Tonographie and gave them a contract to equip complete sound recording studios at Ultraphon in Berlin-Wilhelmsaue.

Among the components for the Ultraphon recording studio Erbslöh and von Willisen made a portable recording system that could be packed into several trunks. With this apparatus they made thousands of records in many European cities for Ultraphon. In 1930 Erbslöh travelled with approximately a hundred boxes of apparatus, blank discs and various auxiliary equipment to Java and the Malay peninsular and over a period of months recorded native music in various locations. Von Willisen continued to improve the reproduction techniques while working in the Ultraphon Studios in Berlin. For judging the musical quality he always had Herbert Grenzebach at his side. The two earned praise for their efforts in recording a Beethoven piano concert at the Musikhochschule, which they recorded simultaneously with the renowned Deutsche Grammophon and for which they received higher acclaim.

It was a busy time for the two, who obtained a substantial income in addition to acknowledgement and satisfaction. In their mid-twenties the spirit of entrepreneurs stirred in them. This showed up more strongly in Erbslöh than in von Willisen, whose thoughts were tied to experimental electronics. Frequent meetings with Dr H E Hollmann of the Heinrich Hertz Institute increased his interest in high frequencies and formed a lasting friendship from which an important professional collaboration was to arise. Hollmann was devoting himself to ultra-short waves and had built in the physics department of the Technische Hochschule in Darmstadt the first transmitter and receiver for centimetre and decimetre waves. In 1928 he graduated with a thesis about Barkhausen electron oscillations.

In 1931 Erbslöh and von Willisen acquired Tonographie and gained financial independence for it by making recordings for brand-name record companies. They had quickly recognized what was remunerative in the business and were soon working not just for record companies. They developed ideas of their own and soon advertising records and political party speeches were part of their business.

Tonographie hit its stride on Forkenbeckstrasse in Berlin-Schmargendorf. There was no time now for their own development and construction of recording equipment. They bought these items from speciality companies that had come onto the market, and von Willisen, acting under the influence of Dr Hollmann, pushed the development of radio techniques. Since telephoning was his favourite method of carrying on business, he developed with technicians from the Neumann Company a wireless telephone on wavelengths below a metre following Hollmann's plans. Owing to the high commitments of Tonographie, these sets were made initially only in limited quantities: the economic gain from this development was to come later.

Von Willisen understood wonderfully the acquisition and manipulation of business connections. The Central Office of the Reichs Post and the Heinrich Hertz Institute advised and supported him. He nurtured with particular care his relationship to the inventors of talking movies, in Germany, Engel, Massol and Vogt, and another talking-picture pioneer, Breusing, belonged to his circle of acquaintances. Georg Neumann, with his enormous range of experience and knowledge in matters of sound techniques, remained a highly esteemed colleague.

Somehow in 1932 a connection was established with the Army Ordnance Department on Jebenstrasse in Berlin. Phonograph records of muzzle and projectile bursts were needed to train troops to measure time intervals and use them to determine locations of enemy weapons. Erbslöh and von Willisen entered personally into this assignment, placing their portable equipment in trenches at the Kummersdorf practice range, and obtained results that satisfied the customer.

At the end of 1932 a friend of the von Willisen family arranged a contact with the NVA through Naval Captain Kieseritzky. He and the NVA leader, Captain Kienast, had the problem of making records to train personnel to recognize the sounds of various vessels, especially U-boats, that were heard with the so-called 'sea-bed stand'. At the centre of this multi-legged stand, which was lowered to the ocean floor from an outrigger, was an underwater microphone. It was connected to the ship by a cable where an amplifier allowed the signals to be heard. Kienast found Erbslöh and von Willisen trustworthy and assigned them the task, which they entered into personally. They adapted their portable equipment to the underwater microphone of the hearing apparatus and over several weeks recorded the noises generated by various ships with a quality never before encountered, filling the NVA personnel with enthusiasm.

Dr Kühnhold, scientific member and leader of the NVA acoustic division, was present at some of this work. He stood at the time at an advanced stage of his horizontal-echo method of target location by means of underwater sound and needed a sound generator of high power. Erbslöh and von Willisen learned that he needed a sound-pulse generator in the 10–20 kHz band with a power in the range of kilowatts. Von Willisen discussed the problem with his friends at the Heinrich Hertz Institute and with his confidant, Georg Neumann. He soon told Kühnhold that he could develop and build the desired generator. After details were clarified, Kühnhold, who had just been named scientific leader of the NVA, gave the contract to Tonographie.

Von Willisen developed the generator with the technicians of the Neumann Company through a sub-contract. Two RV24 tubes drove the power amplifier with two RS15 tubes in a push–pull circuit. A capacitor charged to 4 kV was discharged for the pulsed-anode voltage. They quickly determined the best core material for the 20 kHz output transformer, and by mid-1933 they had built their first pulsed-sound generator. It satisfied the requirements of the NVA and was immediately accepted. Now Kühnhold could take his experiments with horizontal-sound location far enough to produce a provisional report about the standard location apparatus, although the method of completing the apparatus was still open. Neither the 'red-light method', using neon bulbs with the Atlas fathometer, nor the NVA indicator technique, which used a mirror galvanometer that reflected a light beam onto a sliding scale of ground glass or photographic film, gave much hope.

Encountering Tonographie and meeting its owners, Erbslöh and von Willisen, proved to be of great significance to Kühnhold. In the realization of his progressive ideas he had met with little enthusiasm or support from the firms he had heretofore dealt with, but in these two dynamic engineers he recognized men who went at a new problem without reserve or shyness and with plenty of enthusiasm. He recognized in them two entrepreneurs for whom nothing seemed impossible, who only wanted to try.

CHAPTER 3

AN IRRESOLUTE BEGINNING

The conditions within the NVA in the fall of 1933 were characteristic of what followed. Kühnhold had brought forth with his acoustic group a meritorious accomplishment in conducting the research and development that led to a working horizontal-sound locating system. He led a group of scientists, engineers and technicians of about his age, and who respected him as their chief. The extensive physical knowledge from which he drew his ideas made working with him a pleasure. He knew how to motivate and inspire his co-workers, even when the steadfastness with which he carried others along with him in pursuing solutions could be unpleasant.

The immobility of the NVA bureaucracy set limits to his impulsiveness. It was always necessary to convince his superiors in detail of his plans in order to obtain their support and goodwill. Thanks to the unexpected, spontaneous readiness of Tonographie to deliver to him a pulsed-sound generator, which had not existed and thus had to be developed, in 1933 he could demonstrate his sound-location equipment to the Torpedo-Versuchsanstalt (Torpedo Research Establishment, TVA), which convinced them and caused them to provide the support for the further development of the project.

Tonographie received orders to develop pulsed-sound generators for a number of frequencies and placed some of them into production. They were able to increase the generator power to almost 5 kW and could match the output impedances of the generators to the transducers, even transducer groups from different manufacturers. NVA gave Tonographie the task of producing prototypes of sound-ranging equipment according to their own design.

It lay in the nature of things that Tonographie did not have to be bound by fixed contracts with tight limits, and continuing discussions held by Erbslöh, and especially by von Willisen, sufficed for Tonographie. The mental attitude of the two was an excellent basis for an agreeable course for these discussions—both thought on the same wavelengths as their associate at the NVA. That their friend and advisor, Commander (retired) Hermann Eduard von Simson, sometimes took part in the discussions with the NVA and senior service personnel was a definite help.

The collaboration between the NVA and Tonographie had become extremely close by the end of 1933, supported on one side by Kühnhold and his co-workers and on the other by Erbslöh and von Willisen, although one could not speak of a business partnership. NVA disposed at that time of very modest funds for procurement, and the award of contracts to companies fell under the strictest rules for purchases. Fortunately, Tonographie had such a continuing, profitable recording business that they could work for NVA and still turn a profit, even if a contract was not awarded and payment was long delayed.

The good and trusting relationship of the NVA with the owners of Tonographie was noticeable when Kühnhold initiated Erbslöh and von Willisen into his thoughts in fall 1933 about the application of radio waves for target location, an occurrence that took place accidentally.

Von Willisen had demonstrated to the NVA people a Tonographie radio telephone for 95 cm that was intended to allow generally secure communications between warships as a replacement for light signals, a method that would be useful even in bad weather. It awoke in Kühnhold an entirely different interest, however.

Kühnhold told von Willisen of his unsuccessful attempts to observe reflections from ships with very short waves. He had used the Pintsch equipment over the very same stretch of water that had just been used to demonstrate the Tonographie 95 cm radio. He also told von Willisen of his unprofitable discussions with qualified experts, who had assured him that there was no hope for such a technique, given the current status of vacuum tubes. He foresaw high power and narrow beams as absolutely essential for producing usable reflections from metal surfaces at great distance.

Owing to the impression made by the 95 cm radio, he asked von Willisen if he could use the equipment for reflection experiments. Von Willisen agreed immediately and promised to have antennas made in the Berlin workshop that would provide a sufficiently narrow beam.

Tonographie had attempted since mid-1933 with the advice of Hollmann and Dr Theodor Schultes, who worked voluntarily with them, to increase the range of the 95 cm equipment. Originally this apparatus was equipped with a parabolic antenna, jokingly called a 'whisper bag'. When equipped with a Barkhausen tube at both the receiver and transmitter antenna, this was the same technique Hollmann had demonstrated at the 1932 radio exhibition in Berlin. At Potsdam-Werder von Willisen had made range experiments with this equipment many times with Erbslöh, using a variety of antennas.

Coincidentally von Willisen learned at the beginning of November 1933, as the preparations with the 95 cm equipment for these reflection experiments were under way, of a new kind of magnetic-field tube for the generation of centimetre waves at high power. Philips in Holland intended to market these tubes and provided guidance for their application in high-frequency generators. Von Willisen ordered some prototypes of these new tubes immediately, using a middleman from his circle of friends to make the purchase, as they preferred that Philips not know the real purchaser.

When Kühnhold learned of these tubes and their procurement, he was greatly excited, for this opened the way for the construction of a transmitter for his reflection experiments. So the year 1933 ended with Kühnhold involving Tonographie in more activities than horizontal, underwater sound location. He had great hopes that the owners would rapidly confirm his ideas for radio location. For their part Erbslöh and von Willisen were spurred on by the trust placed in them by Kühnhold.

The delivery of a magnetron transmitter with an associated receiver as well as antennas was already completed in 1933. Kühnhold obtained from the chief of TVA, Admiral Hirth, quick financial support for the NVA out of funds from his own department. Those involved knew that Tonographie would help Kühnhold quickly to his goals by avoiding long, bureaucratic delays. He was confident of an early success. Erbslöh and von Willisen became concerned about the unexpected costs, but they nevertheless placed their resources at Kühnhold's disposal.

CHAPTER 4

THE INCORPORATION OF GEMA

The surprising beginning of intensive business relations with the Navy began to cause the two owners sleepless nights. Their firm, Tonographie, was progressing beautifully and enjoyed increasing respect for their products, but the direction that their activities in ultra-short wave radio would take was not clear. In discussions with officials of the Reichs Post, von Willisen had pushed for the development of a radio link with the postal network, even though this was an area already occupied by well settled companies. The thought of applying ever shorter wavelengths and ever smaller antennas for long distance traffic had lodged deep in his mind, and he was further excited by his contacts with Hollmann and the personnel at the Heinrich Hertz Institute. That Tonographie could hold its place among these other companies came about because all were entering a completely new region where experience had first to be collected.

The uncomplicated development and the rapid delivery of pulsed-sound generators of high power was the reason why the naval leadership was able to accelerate underwater sound location. They pressed for tests and further development that would lead to deployable apparatus, and their trust of Erbslöh and von Willisen certainly made Tonographie an obvious and convenient choice for both prototypes and production. Such business possibilities quickened the heartbeats of the two owners. Because they were primarily engineers, and because the authorities for conferring contracts were not well informed, they felt somewhat restrained. What they saw coming seemed almost stifling.

The year 1933 was filled with continuing discussions that Erbslöh and von Willisen held with their patent attorney, Dr Fritz Walter, and with Georg Neumann and his partner Erich Rickmann. They included the owners and partners of the Neumann Company in these talks because it had furnished shop capacity for the deliveries to the Navy, as Tonographie itself did not have a shop capable of producing the large units. The economic and political conditions that governed Germany at the end of 1933 were such that neither firm was ready to undertake the risk of a large commitment to the Navy. They thus decided to form a new corporation, which would be active primarily in underwater sound and could take on the development and production of their radio-telephone sets. Erbslöh and von Willisen were primarily interested in forming this new company, 'GEMA, Gesellschaft für elektroakustische und mechanische Apparate m.b.H.', as they had greater interest in these activities than Neumann and his associates. They were employed in registering the new company with the lawyers and notary but soon separated themselves from it. The name of the company resulted from the components that made up the sound-location equipment: electro-acoustic transmitters and receivers with extensive mechanical control.

The incorporation and registration in the commercial register of Berlin took place on 16 January 1934 with a capital given as 20 000 Reichsmark (about $4500). The stockholders were Erbslöh, von Willisen, Neumann and Rickmann in equal parts for profit and accountability, but Neumann and Rickmann soon left the firm, with Erbslöh and von Willisen then taking over their shares. According to the commercial register, GEMA concerned itself with the development, production and operation of mechanical, electrical, acoustical, phototechnical and photochemical equipment and devices. The leaders limited the purpose of the firm initially to the construction of prototypes and small production series. The company grew through an ever-increasing series of contracts with the government and expanded from this small beginning to a large concern with 6000 employees by 1945.

The location of the first GEMA factory was at Potsdamerstrasse 122 in Berlin W35. Initially Erbslöh and von Willisen worked there as engineers with two co-workers and the volunteer consultants of Hollmann and Schultes. In their typical engineering office they transferred their ideas to paper and sent orders to co-operating firms. They were naturally concerned about the security of their ideas, the natural consequence of their novelty, so subcontractors were not allowed to have knowledge of what GEMA was doing. The single exception were the owners of the Neumann Company during the time they were shareholders of GEMA. Tonographie would henceforth have to remain independent, but the firm's management was sufficiently competent to allow them the freedom required to follow their new interests.

CHAPTER 5

THE FIRST UNDERWATER SOUND
AND RADAR EQUIPMENT

By the end of 1933 Kühnhold had put together the essential characteristics of an acoustic aiming device for the torpedo arm of the Navy. The invention of the sum–difference method gave him a decided advantage for active sound location using microphones spaced relatively closely. A large number of careful experiments were conducted to determine the velocity of sound in sea water under various conditions. It was now necessary to incorporate these techniques and knowledge into a device that would be useful aboard ship for determining in a simple manner both the direction and range for aiming torpedoes.

The Navy allowed GEMA access to all secret documents. They ordered them to develop the desired equipment in co-operation with Kühnhold's group, as there was still no design for a device for sound location that was simple enough for use on a vessel of war. The sound generator had reached such a level at Tonographie that final construction could begin on it, and underwater sound transducers were contracted from Atlas in Bremen and Elac in Kiel. The NVA laboratory had built the prototype of a twin amplifier for the measurements and had tried it out with various methods for making the echoes visible, but the method that used a cathode-ray tube (CRT) was absolutely rejected, as they feared that this tube would be damaged by the shock of a vessel's gunfire. Many techniques using mirror galvanometers with sliding scales for time bases were carried out with the co-operation of Tonographie, but a useful answer was not forthcoming. The enforced solution eventually came from a semi-transparent rotatable mirror in the GEMA laboratories from ideas put forward by Erbslöh and Golde.

Parallel to the work in sound location that was now to be advanced by GEMA, Kühnhold brought life to the experiments for a similar location method using radio waves. He had immersed himself so deeply in the theme of underwater sound as a means of location that radio location simply seemed the logical extension of the earlier work. As a physicist he had no doubt that, by building on the known experiments of Heinrich Hertz, he would reach the desired end, although the experimental basis of observed reflections from transmissions was absent. The failure of his experiments with the Pintsch apparatus on 13.5 cm in early 1933 did not discourage him in the slightest, for he was sure that the power of the transmitter could be raised and the sensitivity of the receiver could be improved so that reflections from ships at great distances would be possible. It was extremely important for him to transfer his ideas from the realm of the impossible to that of the possible.

As soon as Kühnhold had entrusted this problem to Erbslöh and von Willisen he received immediate support from them. Without any contractual agreement GEMA began reflection experiments with the 95 cm equipment in collaboration with NVA. The two had made their reception and range experiments in early 1933 over the stretch of water from the Potsdam Matrosen station on Jungfernsee to the Sakrower church and from the Potsdam War Archives to Bismarck Heights at Werder. Nearly thirty years earlier Slaby had used the same locations to demonstrate the first wireless traffic on considerably lower frequencies. (Adolf Slaby was a radio pioneer and close associate of Georg von Arco, the founder of Telefunken.) They quickly learned that sometimes reception was possible even though there was no optical line of sight between transmitter and receiver, and they explained this as the result of reflection. Inasmuch as this phenomenon was of no use to their work, they assigned no importance to it. Their first priority was to increase substantially the range of their sets.

In February 1934, for the first time von Willisen participated in reflection experiments in Kiel harbour with GEMA equipment. Both transmitter and receiver depended on Barkhausen tubes and, in these experiments, worked on 75 cm. Transmission was made with a three-element Yagi, reception with a simple dipole. These experiments failed to gain clear evidence of reflection from ships. They did observe signals from directions at which the transmitter had not been pointed, but the wide radiation pattern of the dipole did not allow them to say that these came from ships.

In the meantime GEMA had built the first laboratory oscillator with the Philips magnetron. Initially they followed the suggestions of the manufacturer and used two magnetrons connected to Lecher wires, which was not satisfactory, although the 70 W power they generated on 48 cm was encouraging, measured by using carbon-filament lamps as load, some in parallel to reduce the inductance. There were no measuring instruments for decimetre waves on the market, so they had to improvise their own. The wavelength was measured by the maxima and minima on Lecher conductors. The joy at their first magnetron oscillator was seriously muted because it was very unstable and difficult to keep oscillating. Attempts to modulate it by means of the magnetic field and the anode voltage proved most disappointing. Von Willisen spent many hours trying to optimize the anode voltage and magnetic field in order to get the circuit to oscillate even halfway satisfactorily. In the meantime Hollmann set up a Barkhausen receiver for the wavelength of the magnetron transmitter. His first experiments in Berlin used a four-element Yagi for the transmitter and a dipole for the receiver. They succeeded despite all kinds of difficulties.

In March 1934 GEMA installed this laboratory model on the balcony of the NVA building in Kiel. On 20 March 1934, after many preliminary tests, von Willisen aimed his transmitter antenna at the old battleship *Hessen*, which lay 500 m away, but the search for reflections failed, no matter how the receiver dipole was rotated, as the direct transmitter signal overwhelmed the receiver, making it insensitive to the much weaker reflected signals. Replacing the transmitter Yagi with a parabolic reflector and screening the receiver with a large piece of sheet metal did not help. Von Willisen transferred the small, battery operated receiver to a boat and moved in the direction of the old warship. Close to the ship reflections could be observed from the freeboard of the hull, but only if the receiver dipole was shielded from the direct beam of the transmitter. The problem then was obviously how the direct radiation could be suppressed.

Independent of the reflection experiments, the work with the 48 cm equipment took on a completely new form, as the magnetron transmitter greatly increased the range. Also a new oscillator circuit that used only one magnetron yielded much better stability and worked much more reliably than with two magnetrons. **Modulation was a difficult problem, the result as they later learned of the pulling of the frequency. Hollmann had improved the receiver by combining two Barkhausen tubes in push–pull, which was much superior to a single-tube audion. It could be tuned entirely by adjusting the tube's grid voltage, and further tuning elements were not necessary. The receiver could thus be mounted on a tower and tuned from a distance.** The magnetron transmitter and the improved receiver were made into a stationary radio-telephone system and given over to the Navy and the Reichs Post for testing.

While people in the GEMA laboratories were trying to shield the receiver from the direct waves to improve the reflection results, Kühnhold resumed work with the Pintsch 13.5 cm system during 7–12 May 1934. He succeeded in raising the transmitter power to 0.3 W and connected both units to parabolic reflecting antennas with diameters of six wavelengths. He tested this model at Schilksee near the mouth of the Kieler Bucht. The target of his reflection experiments was the 400 ton NVA research boat *Grille*. Reflections were clearly observed from the boat at a distance of 2100 m. The crucial difference here came from the motion of the target ship, which caused the reflected radiation to be shifted slightly in frequency and made the interference of direct and reflected waves into variations in amplitude of the receiver output. Reflections from stationary targets caused no such easily observed variation.

This result was quickly made known to GEMA and unleashed the useful forces of competition for prestige and for the favour of Dr Kühnhold.

Von Willisen himself conducted additional experiments at NVA from 18–28 June 1934, earlier than had been planned. In order to reconstruct the Pintsch apparatus experiments, he built the new GEMA apparatus at the same position at

Schilksee where the successful rig had stood. The transmitter provided 40 W on 48 cm from a parabolic antenna of five wavelengths diameter. The receiver stood on an 8 m high platform, had as an antenna a dipole array called the 'Tannenbaum' and used Barkhausen tubes in the well-established push–pull configuration. The Tannenbaum antenna was constructed following a suggestion by Dr Schultes. In front of a screen reflector were placed four vertically stacked dipoles, whose advantage lay in the ease with which they could be matched to the receiver input. Although the transmitter and receiver were well shielded from one another, the transmitter, whose anode voltage was modulated with 1000 Hz, still leaked into the receiver. This time with the transmitter and receiver standing side by side, reflections from a sheet metal plate that had been erected on a small boat were detected with the rotatable receiver antenna at a distance of 300 m, but the Pintsch arrangement with a thousand times less power observed reflections from 600 m. When a large steamer passed through their beams the Pintsch apparatus detected it at 4000 m, the GEMA only at 2000 m. In contrast to the GEMA rig, the Pintsch showed a pronounced dependence on the amplitude of the received signal and the range to the ship.

The results of these experiments were not satisfactory but were certainly encouraging. Kühnhold saw his ideas confirmed, and the two GEMA partners were excited by curiosity and delight for the project. They recognized that to remain competitive in this new field of radio location, whose limits they were beginning to see, they would need the assistance of more capable people because their current staff could not meet these increased demands on their skills. Their company already required competent scientists and engineers to be continually active in research and development. Given the constraints on time they would have to enter into such limited contacts as the company could economically carry. It did not hurt that the business of Tonographie was flourishing, that they held the confidence of the banks, and that they gained as permanent co-workers Drs Schultes and Brömel as well as the two engineers (with degrees) Dünhof and Schöneborn. Hollmann was available for scientific consultations, and a lucky chance brought to them the services of the superb precision mechanic Heinz Röhrig, who with the help of glass blowers turned out to be a specialist of the first class in fabricating new kinds of vacuum tube.

By September 1934 the GEMA laboratory had developed a new kind of receiver and built its prototype. **A decision had been made to use an intermediate frequency, but the push–pull Barkhausen-tube circuit had been retained due to lack of time to design a new input. The 1150 kHz intermediate frequency (IF) was amplified by a broadcast receiver that also demodulated the 1000 Hz found on the IF. The transmitter was outfitted with new split-anode magnetrons from Röhrig that worked more reliably with the IF than the Philips tubes. A new concept was introduced with a CRT from the firm of Leybold & von Ardenne with which they measured the phase shift of the 1000 Hz signal relative to the generator signal and from which they could determine the distance and compare the result against a target of known distance. The**

frequency of the generator could be varied between 1000 and 6000 Hz. Great effort went into shielding the transmitter from the receiver. The degree of shielding attained allowed reflections to be observed even for stationary targets.

For further experiments GEMA had selected with Kühnhold the NVA research grounds on the Lübecker Bucht where they built a 10 m high wooden tower, the so-called 'GEMA tower'. On this tower was a turntable, on which antennas were mounted, and a wooden hut to protect their electronics from wind and weather. From 12 October to 2 November 1934 GEMA tested their newest designs. The research boat *Grille* was at their disposal and was successfully ranged to 7 km. Until then only a rough measurement of the phase shift between the transmitted and received 1000 Hz signals had been made, which did not suffice for a range measurement of satisfactory accuracy because the receiver could not be satisfactorily shielded from the direct signal. Mounting the receiver and its indicator 200 m away from the transmitter with synchronization brought over by wire made a significant improvement. Now delay times calculated from phase shifts agreed with the calibrated values. In these experiments the direction of the reflected signal could be roughly determined by use of the CRT, and with this detached-receiver arrangement the *Grille* could be detected at 12 km. By accident during these studies a strong echo resulted when a Junkers W-34 flying boat passed through the transmitter beam.

These experiments showed that the state of development was indeed rather shabby. Much fundamental knowledge had been obtained, but one could not speak of real measurements. Be that as it may, the series of experiments took on additional significance when members of the influential TVA were present in addition to those of the NVA. Until then Kühnhold had had little or no financial support, indeed only what could be diverted from other projects. This situation was changed by the positive impression made on the TVA group, who made available in a most un-bureaucratic manner for a limited time 70 000 RM (about $17 000) for the continuation of the work, providing a sound financial basis for the first time.

Given this point of view the two GEMA companions, Erbslöh and von Willisen, lost the scepticism and reserve that some had felt. Now they were in a mood to put all, or at least an essential part, of GEMA's resources behind radio location. It was obvious that in going beyond sound-location methods to those of radio meant giving the Navy the ability to see in the dark and therewith an enormous tactical advantage. On applying for patents the inventors discovered to their surprise the earlier work of Christian Hülsmeyer, who patented in 1904 his 'Telemobiloskop' for locating ships. The searches brought to light, however, no evidence of a known method of determining direction and range by means of active radio location. From these preliminaries Erbslöh and von Willisen learned that some of their claims were fundamental and that GEMA could secure exclusive rights in this field. From discussions that Kühnhold had held with Dr Wilhelm Runge about radio location with decimetre waves, they knew that since 1933 Telefunken's current research had seen such frequencies useful only for communication. GEMA formed a link with the firm of C Lorenz for the acquisition of tubes, and Lorenz was also interested in decimetre waves for communication. These common interests brought about exchanges of ideas and some relaxed collaboration. From mid-1935 Lorenz showed great interest in GEMA's work in radio location. This was at a time when GEMA had tested their first sets together with NVA.

Almost fifty open and secret patent applications allow one to recognize the extent to which the company had progressed in sound and radio location, all in close association with NVA. The newly gained knowledge necessarily pointed to specific applications. They were obliged to deliver to TVA a functioning set equal to or exceeding the accuracy of its optical systems, and TVA alone had funds for the development of new techniques.

After their first successes Erbslöh and von Willisen began to think, with the memory of the *Titanic* disaster in mind, of a device that could not only see in a given direction but see all around. Erbslöh in particular gave the first impetus for a panoramic radar, but first they had to satisfy the demands for naval weapons requirements. This was not only the ability to discover a target in the dark but also to determine its bearing and range accurately, according to later German terminology a 'Funkmess Gerät' (radio-measurement device), a name coined by Kühnhold in September 1935. After 1945 the American term 'radar' (the acronym for 'radio detecting and ranging') began to replace 'Funkmess' in German usage.

The results of the October 1934 experiments at Pelzerhaken had shown that the continuous-wave methods used until then had no future. Everyone involved was agreed that simultaneous transmission and reception would give no useful system. The overwhelming of the receiver by the direct transmitter signal had to be avoided, so the receiver had to be turned off for the time of the transmission and then returned to full sensitivity in order to record the echo. This was especially true when it became known that the transmitter power had to be greatly increased in order to extend the effective range of the set.

It was on the recommendation of Hollmann and Schultes that GEMA applied the pulse technique that was already being used in ionospheric soundings for the propagation of middle and short-length waves as well as for their sonar techniques. Hollmann had had such experience in 1930 as part of his work at the Heinrich Hertz Institute in preparing for the 1932–33 International Polar Year. The German contribution was to consist of equipment for sounding the ionosphere with a short-wave transmitter that was modulated periodically with microsecond pulses. Schultes undertook this task and immediately began developing a way of modulating a magnetron transmitter with very short pulses. In order to protect the receiver he made experiments with neon lamps at the receiver input, which the transmitter pulse would fire and thereby short-circuit the input. At the beginning of the time-delay measurement, the transmitter pulse triggered a sweep voltage for the CRT, thereby providing a deflection proportional to time. The receiver output voltage of the echo signal was applied to the other pair of CRT deflection plates, which allowed the time to be measured from the known sweep-speed.

This form of presenting the data was rejected by the Navy because of the presumed fragility of the CRT, the same objection they had for its application to sound location, but practicality compelled GEMA to retain the method, at least for the present. Partly in order to circumvent objections to the CRT but more because of the desire to overcome the restrictions on time resolution imposed by the tubes available at the time, which only had 200 mm screen diameters for a linear or circular trace, GEMA initiated the invention of their adjustable range-measuring delay circuit, subsequently called 'Messkette', readable from a control dial or numbered wheel. This allowed the delayed pulse to be matched with an accurately generated marker pulse. To counter the Navy's objection to the CRT, they used a much smaller, less fragile version as the indicator. The task of designing and developing the first Messkette was undertaken by Professor Salinger of the Heinrich Hertz Institute.

Röhrl, who as a NVA staff scientist particularly concerned himself with magnetrons, and his colleague Rindfleisch were often active as guest workers at GEMA. It was with Rindfleisch's co-operation that the method of lobe switching was developed and tested for determining the direction of the echoes. This was a method that had similarities to the sum–difference technique of sound location. By comparing the receiver outputs of two antennas whose radiation patterns overlapped in a specified manner, one could obtain significantly higher directional accuracy than with a single antenna that depended on determining the direction of maximum signal. In January 1935 GEMA began experiments in which each of two antennas was connected to its own receiver. Because it was difficult to maintain equal gain on the two receivers, a motorized rotating switch that fed a single receiver was preferred, and the comparison of the two antenna voltages was done with a stereoscopic optical arrangement for viewing two CRTs, later replaced with a single CRT. With this system bearings were determined with an accuracy ten times that obtained with the maximized signal of a single antenna. The results were soon to be better than $0.2°$.

Since January 1935 GEMA had utilized the RCA 955 'acorn' tubes for its magnetron radio telephones. They operated in the super-regenerative mode with quenching, which was done by a separate quenching oscillator. It proved very good in receiving magnetron transmissions that had a high degree of frequency modulation, although the typical noise of the quenching frequency was troublesome.

This new receiver was tried at Pelzerhaken especially for locating aircraft. An antenna that could also be rotated vertically for the audio-modulated magnetron transmitter was mounted on the turntable of the GEMA tower. Von Willisen ordered a Junkers W-34 flying boat to fly back and forth in front of the antenna, which was moved horizontally and vertically. The receiver was built into a coach for which a suitable position was found behind the transmitter. The demodulated receiver IF signal showed a strong echo from the aircraft, but at more than 12 km distance the strength of the echo dropped so far that it was no longer distinguishable from the direct transmitter radiation. This proved that leakage from the transmitter seriously limited the maximum attainable range for a continuous-wave (CW) system.

By the end of February 1935 GEMA terminated experiments with audio-modulated CW transmitters in all its various combinations of receivers and indicators for ranging and dismantled all their equipment at Pelzerhaken. New pulse-modulated equipment with CRTs that might deservedly be called by the name radar took shape in the Berlin workshops.

A transmitter for 52 cm that used a split-anode magnetron of their own construction and to which a rectangular 2 μs high-voltage pulse was applied yielded output pulses of 1 kW peak power at 2000 Hz repetition rate. The receiver was a one-conversion super-heterodyne using 955 acorn tubes at the input. The Q of the tuned circuit was increased by a feedback-controlled mixer stage, a notable GEMA design. The 7 MHz IF amplifier had four stages of AF7 tubes and had a bandwidth of 200 kHz. The IF output was rectified with AB2s and fed to the vertical plates of the CRT after amplification by AL2s. According to the sine-wave deflection described later, the spot traversed the screen from the centre to the right and from left to the centre, in each case requiring 66 μs. The screen was graduated from 0 to 20 km, a distance considered sufficient, given the circumstances. The modulation and repetition rate was derived from a stable 2000 Hz frequency generator, corresponding to a maximum echo path of 75 km. An uncalibrated phase-shift circuit allowed the transmitter pulse to be adjusted to the zero mark of the indicator scale. During the period of the transmitter pulse, a neon bulb at the input of the receiver burned and greatly reduced the effect of the much larger signal.

GEMA had no experimental ground in Berlin, so they installed their new radar straightaway on the Pelzerhaken tower. For the transmitter antenna they mounted a so-called mattress, a vertically polarized array of ten dipoles placed in front of a reflecting wire mesh. The receiver with indicator was set up away from the transmitter but synchronized with it through a cable. The receiver antenna was a similar dipole array with reflecting screen. Initially, the direction was obtained by rotating the antenna for a maximum signal. The antenna for accurate directing using lobe switching was not yet finished.

Even the first experiments in adjusting the antennas presented them with a completely new picture. From the direction of Neustadter Bucht, a small inlet within the Lübecker Bucht, came echoes without a ship being within the beam. These remarkable echoes proved not to be equipment problems, which were suspected at first, but reflections from a large forest near Scharbeutz at a range of 15 km. This wood served immediately as a fixed target for optimizing the set. From other directions came more or less strong echoes that were identified with hills and buildings.

By the beginning of May 1935 the preparation for observations with the set were completed. In the tests made while setting up the equipment they determined that the blocking of the receiver input with the neon bulb was so effective that the receiver could be operated in its rightful place on the tower next to the transmitter. The receiving antenna was then mounted above that of the transmitter, allowing them to be moved together.

The research boat *Grille*, renamed in the meantime *Welle*, ran back and forth in front of the GEMA tower for the first measurements. Surprisingly the new equipment did not achieve the effective range of the old. The course of the *Welle* could only be followed out to 3 km, although the receding of the echo pip was easily seen on the expanded time base as the vessel moved away. Just as with the Pintsch equipment the amplitude of the echo changed according to the orientation of the boat.

In these experiments the new receiver did not prove at all satisfactory. The pass band seemed insufficiently wide, and the IF had a tendency to oscillate. As a consequence the gain could not be turned to maximum, and this was the cause of the disappointing range data. Additional shielding had no effect, and the experiments were terminated.

The laboratory designed a new, dual-conversion receiver, in which the IF amplifier was given a second stage at a lower frequency. The first IF at 15 MHz used two stages with sharp-cut-off 4673 pentodes feeding an ACH1 converter that produced the 7 MHz of the second IF; a further stage with 4673s amplified the second IF, which was rectified with an AB2; an AL4 output stage amplified the demodulated signal, and an AB2 suppressed the overshooting. Based on the high gain from the sharp-cut-off pentodes, the bandwidth could be increased from 450 to 500 kHz.

In May 1935 Professor Salinger of the Heinrich Hertz Institute delivered the first calculated Messkette intended for exact range measurements out to 20 km. It was built as a homogeneous symmetrical delay line and had, for the present, only 20 steps, each of 1 km. This chain had a design that was generally independent of frequency and allowed one to move the echo pip onto the zero position of the CRT. When this had been done, the scale on the circuit gave a value proportional to the range. From this device GEMA expected, with finer graduations and with careful marking of the flanks of the transmitter and receiver pulses, an accuracy in range comparable with that of optical methods. The path to that goal was to be a long one.

As a result of the short time allowed for the construction of the first Messkette, it had a serious error. The capacitors at the line's input and output had been improperly adjusted and the coils, which were open spools without iron cores, affected each other, in part a consequence of the constrained space that GEMA wanted.

Now began the scientific collaboration of Dr Walter Brandt, who, in addition to other tasks for which he contributed much, developed the Messkette into an instrument of high precision that became typical of GEMA components.

After his graduation in Göttingen where he studied filter theory under Professor Cauer and wrote as his thesis 'Electric wave filters to separate various frequencies into proper channels', Brandt received a staff position at the Heinrich Hertz Institute in Berlin early in 1934. He developed routing circuits for multi-channel high-quality amplifiers and held lectures on filter theory. In accordance with these activities he calculated filters and routing circuits for GEMA, allowing the AF band of their radio telephone to be divided into eight channels. Through a temporal displacement of the channels the voice was to be made unintelligible. This was part of the task given GEMA for secret radio telephony. In awarding the contract the Navy wanted fleet command and ship commanders to be able to discuss matters securely and without being overheard.

Because of complaints about the first Messketten, the Institute transferred Brandt to GEMA as a consultant. **He re-designed the circuit from scratch with 20 steps of 50 m and 19 steps of 1 km for the 2000 Hz repetition rate. For coils he used the toroidal Pupin forms found in long-distance cables that were made by AEG's Kabelwerk Oberspree, as he could pack these closely without magnetic shielding. The chain remained electrically closed and was terminated by a manganin resistor. The switching of the coarse steps was done by connecting them step by step to the deflection amplifier. A multi-tapped coil at the input determined the fine steps.**

An improved radar set with Messkette now stood on the GEMA tower at Pelzerhaken where it was aimed at the Scharbeutz wood, tested and adjusted. Von Willisen became its permanent servant, who as representative and co-ordinator for GEMA had earned significant weight with the NVA. The naval life appealed to him, and it touched his youthful dream of being a radio officer. He fulfilled the dream a few years later when he entered active duty as a communications lieutenant at the Flensburg-Mürwick school. He also passionately loved sailing and was always, if the opportunity permitted, on sailing trips with his friend Lieutenant Rath, commander of the *Welle*.

Fortuitously in July 1935 the cruiser *Königsberg* dropped anchor in the Lübeker Bucht about 4 km from Pelzerhaken and produced a strong echo in the set on the GEMA tower. This excited interest as to how far this large ship could be tracked. Lieutenant Rath used a trick to induce his comrade on the bridge of the cruiser to execute various manoeuvres within the radiation pattern of the radar. Rath claimed they were testing a secret device that indicated which boilers of the vessel were being fired, as the real reason had to be kept from the ship's commander. It all went off to general satisfaction.

During these tests for the first time they performed experiments that compared the signal strength with two dipole groups as antennas for two receivers. The output of one receiver was inverted relative to the other, which gave positive and negative echo deflections on the time base of the CRT. With a stereoscopic display the operator could adjust the two pips so as to equalize their amplitudes by rotating the antenna assembly, which had the axes of the two pointing in slightly different directions. The method showed advantages over determining the direction from maximum signal. That the bearing accuracy during the experiment was only moderate resulted from mechanical problems that the turntable still had. The play was too great for the later-attained 0.1°. The manipulation and adjustment of the two receiver channels gave von Willisen and Rindfleisch, who operated the set, significant difficulty, and they did not attain exceptional range on the cruiser. Nevertheless, they tracked the vessel to 8 km so accurately by swinging the antennas back and forth that the mechanics of the turntable set the limit.

Kühnhold now had the courage to demonstrate the current stage of development to Navy Command, as his intention for further work in radio location would require a regular contract for GEMA. TVA had covered some of the costs with funds diverted from other projects, but GEMA's expenditures were growing substantially in terms of personnel and material, and their business office had become moderately interested in what the future of their radar work with the Navy would bring.

Final developments of the active sound-location equipment were carried out by GEMA almost hidden amidst the work of the radar section. Although the stormy advance of radar had pushed the acoustic work into the background, it was only after Brandt, at first a volunteer collaborator, came full-time to GEMA that the remaining development of an indicator with twin amplifiers for the sound-location equipment advanced beyond the NVA preliminary work. At the start of 1936 Brandt undertook, as well as the Messkette, the further development of a twin-amplifier prototype, which Kühnhold had used for some of his experiments. A new idea for employing the CRT in the indicator was proposed by Erbslöh, which GEMA presented to the NVA, but as before the Navy did not want to consider the use of CRTs aboard warships. Perseverance and demonstrations that, with proper mounting, these tubes could withstand the shock of the firing of large calibre guns finally gained GEMA the desired change, and CRTs were finally permitted for sound and radio-location equipment aboard warships.

CHAPTER 6

DETE- AND S-EQUIPMENT

On the basis of a wise decision that Erbslöh and von Willisen had made, GEMA had produced in the summer of 1935 its first radar, which they designated EM-2, the term EM being derived from 'Entfernungsmessgerät' with the 2 identifying radar and a 1 designating sound location.

In addition to achieving Kühnhold's idea for radio location quickly, the two entrepreneurs had other, important reasons for binding themselves to this new technology. Through the hiring of more than twenty full- and part-time employees, and with intensive support of the NVA, their firm was strongly committed to the target location method, which they assumed was not unique just to their own country. From day to day they made inventions of devices and of techniques that they covered by open and secret patents. A competent attorney took care of securing their rights and at the same time kept them informed by inquiries and comparisons with what was going on elsewhere at home and abroad.

The business relationship with Lorenz required that they also be informed, at least in part, about the reflection experiments that GEMA had carried out with NVA. Lorenz had attempted without success to obtain a direct path to the NVA location projects and their own exploration of radio telephony on decimetre waves had led to a certain degree of rivalry with GEMA, which restrained collaborative efforts. Erbslöh and von Willisen were afraid of a marriage of convenience with Lorenz because they thought their dowry too small, for as Lorenz's interest in radar showed, they expected that GEMA would come forward with a complete radar design.

The planned demonstration of a radar set to the Naval Command was not to be made with provisional and lashed-up laboratory equipment, and GEMA, as the developing firm for active radio location, was determined to make a lasting impression. They had to gain trust with the service centres that decided such things in order to become the exclusive development enterprise for the expanding Navy for this new weapon. Kühnhold, as the spiritual father of radar, was strongly concerned that his ideas be properly presented to the naval leadership, and above all he wanted GEMA's activities to be secured by contractual arrangement. Until then GEMA was the chosen workshop of the NVA, and the Navy must be induced to provide the means for the continuation of this work at GEMA.

Although a contractual agreement for compensating GEMA for its accomplishments still lay in the distant future, GEMA's management did not save on the costs for producing the components of the three exhibition radars. In day and night shifts they demanded construction that bespoke the impression of professional mastery. Erbslöh and von Willisen inspired their workers by working with their craftsmen in order to have at least one radar ready for demonstration.

They developed a new transmitter, the so-called 'Torpedo transmitter', with 1.5 kW pulse power from a GEMA magnetron that was mounted directly at the antenna. The receiver was a double super-heterodyne with 15 and 7 MHz IF and a bandwidth of 400 kHz. Two receivers with neon lamp input protection were planned for lobe switching using inverted signals. The indicator was equipped with CRTs of 180 mm screen diameter, which were viewed with inverting mirrors. The maximum range was 20 km and the Messkette was appropriately dimensioned. The components were mounted in a covered rack that formed the complete unit.

In early 1935 a new turntable was mounted on the GEMA tower at Pelzerhaken. The principal part of this arrangement for setting angles accurately was the turntable of an optical rangefinder, which TVA provided. For further experiments GEMA had a series of dipole arrays with reflecting screens made that could be connected in horizontal and vertical combinations. Below, on the stump of the turntable, two of these arrays were mounted next to one another and connected to the transmitter behind them. Two receiver arrays were mounted above them and so inclined relative to one another that their radiation patterns crossed by a few degrees to the left and right those of the transmitter. To carry out lobe switching they were connected to two receivers. A continually running calibration generator was incorporated to ensure that the two receivers remained electrically equal.

The first radar of this design was completed in August 1935 in GEMA's workshop and then mounted on the Pelzerhaken tower. Von Willisen personally supervised the construction and undertook the adjustment and testing. The wood near Scharbeutz again proved its worth as a fixed target. Throughout the adjustment, research boats were brought in to try out and optimize the lobe switching. The use of two receiver channels again proved a problem because their gains and frequencies had to be constantly checked and adjusted. Nevertheless the high angular accuracy attainable with this method was bewitching.

By the beginning of September everything functioned so well that Kühnhold invited the Navy Command to a demonstration, and 26 September 1935 would be the day on which much would be changed and decided.

Besides the Commander of the Navy, Admiral Raeder, there were at the demonstration Chief of the Fleet, Admiral Carls, and Chief of Naval Ordnance, Admiral Witzell, as well as high-ranking officials from various bureaux. There

34

was even a noticeable positive reaction as the guests from the military and technical leadership from Kiel were taken to the GEMA tower. The appearance of the 'mattresses', a form of antenna completely new to one of the senior officials from Berlin, made a strong impression, which became stronger when they did not encounter the expected jumble of laboratory wire but a radar set that, although not a piece of equipment ready for a warship, had to be seen as technically advanced. Erbslöh and von Willisen's calculation, that the first impression is often decisive, paid off completely. Only later was it realized how important that inspection was to the subsequent behaviour of the men from Naval Ordnance. They were less interested in appreciating Kühnhold's work than in seeing how the new location technique worked and how it could be used as a weapon.

The research boat *Welle* and the larger artillery school boat *Bremse* stood ready to be tracked. At the base of the tower an optical rangefinder was set up for comparison measurements, and GEMA radio telephones allowed communication between tower and boats.

On the tower von Willisen served as radar operator supported by Rindfleish and a few other NVA personnel, and they were able to demonstrate radio location in a convincing manner. The *Welle* was tracked to 7 km, *Bremse* to 8.5 km, distances that were shorter than had been attained in earlier trials and traceable to the 'squinting' of the two receiving antennas. A series of range measurements were found to be within a hundred metres of the values from the optical instruments. Especially impressive was the series of bearing measurements to an accuracy of $0.1°$. While von Willisen held the receiver adjustments to maximum performance, Rindfleisch moved the antenna assembly to obtain the exact direction. Large lenses, installed on the indicator screens before the tests, greatly aided the equalization of the receiver output signals. Viewing stereoscopically was of little help.

This official demonstration was without doubt a great success for Kühnhold and the NVA. Hardly less was the success of GEMA, whose two leaders showed themselves to such advantage to the Naval Command. During the inspection an excited discussion arose as to what possibilities the new technique opened up for the Navy. At the end came the decision of the Naval Command that further development and testing was to be driven forward as vital. This decision touched not only radar; GEMA's continued development of sound location was to be promoted and accelerated.

The successful representation of GEMA as a development firm at the side of NVA strengthened Kühnhold's position relative to his immediate superiors, so it was proposed that the company become the exclusive contractor for radio- and sound-location technology. Naval Ordnance would take care of the contractual details.

An essential point for the Naval Command as a result of the demonstration was the question of whether and to what degree radar would bring strategic and tactical advantages to the Navy relative to other nations and how these advantages were to be protected. Secrecy was immediately imposed not only on naval units but also on GEMA. Erbslöh and von Willisen could state convincingly for their company and the circle of co-workers involved that the requirement of secrecy as required by law could be quickly enforced. They were forbidden to use the designations EM-1 and EM-2 any further. A joint exploration of possibilities led to 'DeTe-Gerät' as a cover name for the radar, as it might bring the uninitiated to think that the meaning was 'Dezimeter-Telegraphie-Gerät' (decimetre wireless). This was held to be harmless because GEMA, apart from radar, already produced decimetre wavelength radios. No codename was immediately found for the sound-location equipment. It became simply a 'Sondergerät', S-Gerät, hereinafter 'sonar'.

Naval Ordnance found radar a valuable augmentation of their existing optical equipment. They realized that it provided a means of directing the fire of ships' artillery even in darkness and fog. A range of 20 km was seen as sufficient for artillery, so the creation of sets with this capability was recommended. But Kühnhold saw further. He saw the possibility of ranges extending to 75 km, corresponding to a pulse-repetition rate of 2000 Hz, given higher transmitter power and more sensitive receivers. This would give radar a great tactical significance in being able to survey the entire scene of combat.

Kühnhold always found understanding and support for his wise thoughts concerning the future with Erbslöh and von Willisen. These three men enhanced one another, one can say almost ideally, in the harmonious way in which they worked, and this good understanding carried over to the co-workers on both sides. Kühnhold abhorred every form of bureaucracy and the formality of officialdom, impressing him all the more by the generosity with which he was served by GEMA. He understood well enough, however, that this altruism had its limits, and that the economics of a corporation could not be ignored and had to be secured. Erbslöh and von Willisen were not reserved in pointing out to him that their solicitude for the Navy eventually had to be remunerated. They also wanted in any case to avoid being economically dependent on the Navy's manoeuvrings. Many possibilities were evident to them, as the ability to see in the dark had far more possible uses than just another technique of war. Their preliminary accomplishments had certainly paid off. With the presentation of their radar technology, even if only known to the Navy, they had made a great advance over other companies.

Erbslöh and von Willisen were courageous enough to propose to the Naval Command to have the final forms of DeTe- and S-Geräte ready for service in 1938. In a countermove the Navy let it be known that, in view of this schedule, they would leave the entire work to GEMA and confer on them alone the relevant contracts. It would be foolish to suppose that the two suggested that date frivolously. By then they could build on their experience and the enthusiasm of their co-workers. Even if old established technical firms were composed of more research personnel, the young GEMA with Kühnhold's support and its young co-workers had gained an exceptional position in sound and radio location.

At the request of the fleet, and especially Naval Ordnance, GEMA dismounted the radar on the Pelzerhaken tower and installed it temporarily aboard the *Welle*.

These naval groups were, however, not in full agreement about the use of the new device, and wanted to have its utility aboard ship established.

To begin with, the set aboard the *Welle* would be tested for its range against sea and land targets during a trip to the dock in Kiel. An array of four vertical dipoles for transmitter and receiver was placed near the optical rangefinder on the signal deck; the transmitter and modulator stood with the receiver and indicator in a rack in the chart room near the bridge. For the present they relinquished the accurate determination of bearing using lobe switching and of range using the Messkette.

During the early days of November 1935 they located ships continuously, both large and small, on the trip to Kiel, which was specifically for experimentation and which was extended in time. They attained ranges of 6–7 km for large vessels and 2 km for tugs. Targets on land, for example large buildings on the island Fehmarn, were seen on the indicator at distances of 20 km.

During this trip, during which von Willisen took part, it was definitely established that radio location worked just as well at sea as on land. With the understanding and co-operation of the captain manoeuvres to avoid collision were undertaken using the radar indicator, although the captain insisted these exercises were to take place only under conditions that also allowed sight. Before *Welle* was made fast to the dock at Kiel she sailed for further trials on the ford, the results of which determined that a permanent radar set should be mounted in this ship for continued experiments. Until a new set could be installed, one which would be improved over the demonstration set, the signal deck should be enclosed as an experiment room, and a rotatable shaft installed to hold the antennas.

CHAPTER 7

FIRST SURPRISES

1935 ended for Erbslöh and von Willisen more or less satisfactorily. The incorporation of GEMA had proved to be useful, and the investments did not appear to have been for nothing, although the financial load on Tonographie had at times pushed things to the limit. After the discussion that followed the September demonstrations GEMA received in 1935 prompt, uncomplicated reimbursement for all their earlier uncompensated efforts. The two projects, DeTe- and S-Gerät, were to be pushed, as the fleet leadership had in the meantime deemed them to be 'the most urgent task in the NVA programme'. In October the first negotiations between the official offices responsible for conferring contracts and the GEMA management began. A special office of Naval Ordnance was made responsible for overseeing all contracts with GEMA pertaining to development and delivery. Erbslöh and von Willisen had henceforth to endure the official ways in which contracts were awarded, but they could now be sure that the efforts of GEMA would be suitably valued. The need for new management for the firm was solved by the engagement of Mr Alfred Steinwender, who was well versed in the art of dealing with official offices. The question remained, however, of where the company was going and how it was to evolve.

With these first contracts GEMA received the regulations for the security of secret material, which the owners knew were coming, but as the details were placed before them they realized that they had lost much of the freedom of decision that had marked their original work.

It was only natural that GEMA's two engineers and founders did not yet have the aggressive routines of clever entrepreneurs for carrying out such a big project. They had no intention of expanding the company to one with a large production capability. They wanted primarily to exploit the fields of sound and radio location with scientific co-workers and in collaboration with other firms and organizations. The application of centimetre and decimetre waves to problems of communication and location presented them with problems that they could solve alone or with the friendly support of scientific institutes only with difficulty. They expected to have to take on the help of experienced electronics companies and opened early business relations with Lorenz in Berlin-Tempelhof. This connection had come about through Hollmann, who had ordered Lorenz to fabricate his special Barkhausen tubes in their laboratory. Some of the simple push–pull tubes of this kind, which were incorporated in GEMA transmitters and receivers, were manufactured in the Lorenz laboratory at his instruction. There was a trusting relationship between the two firms, with Erbslöh and von Willisen on the one side and Technical Director Herzog and his co-workers on the other, with a similar collaboration having been developed among the specialists. Notable had been the help both in advice and service by Mr Löpp, an expert in the Lorenz tube laboratory. In trying out decimetre radio telephones the two firms co-operated and exchanged a lot of information.

In maintaining these successful relationships for radar, encompassing GEMA, Kühnhold and the NVA, little was withheld from Lorenz, although they were not informed about some details. In contrast to Telefunken, Lorenz had recognized in 1935 the great importance of the reflection experiments that GEMA had carried out at Kühnhold's request. Lorenz's wish to participate in these experiments was shattered by a prohibition by the NVA, even though it was known to them that Lorenz had business relations with GEMA and had some combined laboratory activities.

The good relations and easy collaboration with Lorenz was completely welcome to Erbslöh and von Willisen, and Lorenz also valued this relationship, as they saw in it possibilities for obtaining a lead for themselves in this new field of radar. This common interest gave rise to hesitant, but later concrete, discussions about establishing relations on a contractual basis. An initial proposal that Erbslöh and von Willisen discussed showed the self-confidence that they had gained from the successful demonstration to the Naval Command. They had won much professional weight and could bring considerable force to their negotiations with Lorenz. The decision to join a strong partner to combine operations was favoured as a means of overcoming the requirements for a significant increase in personnel and financial backing that the expanded business with the Navy brought with it. Lorenz's generous financial proposal was additionally enticing in a situation in which GEMA was compelled to make advances of about 100 000 RM (about $24 000) to the Navy.

On 30 December 1935 GEMA and Lorenz concluded a contract for developmental work. The agreement covered the entire working field of GEMA and covered more than forty patents. It ruled that Lorenz had to pay continuing monthly laboratory contributions in addition to an initial payment. Lorenz was to be, except for a few items, the exclusive production plant for all GEMA equipment. The contract called for common holding of patents and for disclosure of experience and opinion. Common holding of patents was desirable because the development work by GEMA might encroach on the rights of some firms in the concern to which Lorenz belonged.

In April 1936 the Naval Command declared the contractual association of GEMA with Lorenz to be unacceptable. The reason lay in Lorenz's international entanglements. The NVA directed GEMA through Naval Ordnance to suspend operation of the contract immediately and initiate talks with Lorenz to bring about a complete termination. GEMA was faced with an accomplished fact and was astounded that Lorenz, for whatever reason, had agreed to this without serious argument. In the discussions that followed Erbslöh was able to explain the matter so that no tense relationship arose. It was, of course, not possible to retract the knowledge Lorenz had gained from their unimpeded access to GEMA's patent files, knowledge that could now be applied to their own ends.

Another unpleasant surprise followed. Before Hollmann had become a consultant for GEMA he had finished and given to his publisher the manuscript for *Erzeugung ultrakurzer Schwingungen* (generation of ultra-short waves), the first volume of his book, *Physik und Technik der ultrakurzen Wellen*, which was to excite great interest in Germany and abroad. He completed writing the second volume, *Die ultrakurzen Wellen in der Technik*, at the same time that he was stimulating GEMA with his knowledge of ionospheric sounding to persuade them to use decimetre waves for radio location by displaying the echoes on a CRT. Unconstrained, he included his experience in the radio location of ships and aircraft in his manuscript. Under the title 'Seeing with electric waves', he indicated the possibilities of radar, drawing on the work at GEMA as a method for the prevention of collisions of vessels with one another and with icebergs.

On the basis of the new secrecy requirements, Hollmann as a GEMA co-worker had to submit his manuscript to the naval censor, who eliminated much of his text. This hurt him profoundly and violated his sense of freedom of expression. Erbslöh and von Willisen were also shocked by this decision but were unable to have the ban lifted. They came to realize that they now had to submit themselves to rules set by the contract, if they were to remain active in a field so secret as radio location. Hollmann was not prepared to submit to this and quickly terminated his associations with GEMA and founded his own private laboratory for high-frequency technology and electrical medicine.

The draft of the contract regulating the working conditions for GEMA with the Wehrmacht Fiscal Office displeased Erbslöh and von Willisen because it contained conditions for charging GEMA with the development, construction and production of special equipment for the Navy, including apparatus for which special secrecy was imposed. Furthermore GEMA was obliged to apply their knowledge and inventions solely for the Navy. They were forbidden to accept orders from the open market or even from the Army or Luftwaffe (Air Force).

GEMA almost had to produce proof that they would be able to provide devices that would extend and improve significantly on human perception. It was not difficult to recognize what uses their equipment would bring to navigation. The first successes were indeed at hand, but the way to the first units that could be deployed on land or at sea was a long one. Erbslöh and von Willisen had to decide whether they should follow the path of free independent research and development, which was financially uncertain and which, there was some reason to fear, the new state might not allow, or if they should continue developing and producing secret target location technology for the Navy. In the latter case they would be assured of long-term financial support, but their entrepreneurial freedom, which had so characterized their lives until then, would be seriously hindered.

After careful consideration they decided to work with the Navy.

CHAPTER 8

DROP THE MAGNETRON, PICK UP THE TRIODE

The magnetron oscillator generated no great love for itself at GEMA. It was easy enough to fabricate. In principle it was just a diode with an anode split into two cylindrically shaped halves with a tungsten filament concentrically centred. The electrodes were contained in a glass envelope with the anode connections passing out of the top and connected to a Lecher-wire resonant circuit. The key to its operation was, of course, a magnetic field parallel to the filament. An electric field that results from a potential being applied between anode and filament is at right angles to the magnetic field, and the values of these two fields are critical in determining if and how oscillations occur. A variation in the strength of one or the other of these two fields causes an undesired change in the frequency, a deficiency that had frequently provoked von Willisen and had disrupted a number of experiments.

It was not possible with triodes to produce frequencies high enough to allow a sufficiently high concentration of the radiation pattern with directional antennas, if powers of kilowatts were demanded. A competent scientist, Professor Karl Kohl of Erlangen, had in 1933 placed the construction of power tubes for decimetre and centimetre waves well into the future.

The evaluation of the experiments in the summer of 1935 had shown that it was the unstable operation of the magnetron that prevented the expected maximum range of 20 km from being attained consistently, and only by constant tuning of the receiver could respectable distances be observed at all. Through some source von Willisen had obtained some American power tubes for decimetre waves and after experiments with them decided to make a decimetre power triode with help from Röhrig and limited help from Lorenz. Röhrig had for some time tried out constructions relevant to such tubes.

Independently, von Willisen began to design and construct metre-wave apparatus using transmitter tubes obtainable on the open market. In his search for tubes with small input inductance and system capacitance, he came upon the Telefunken RS207 through friends at a Reichs Post short-wave transmitter. Although intended for short waves, it had been found useful for ultra-short waves as well. The grid and the anode were supported on short leads entering at the side of the envelope. With 300 W heater power the cathode seemed to be able to endure high pulse emission.

Next he built a self-excited Huth–Kühn circuit for 2 m waves with two RS207 tubes connected in push–pull. They were pulse modulated with 8 kV anode voltage and generated 15 kW. Both tubes stood next to one another on a chassis; the grid and anode inductances were formed by rings of sheet copper that were connected directly to the respective leads on the sides of the tube envelope. This construction was to remain the model for all later GEMA transmitters for tubes that had grid and anode connections on the sides of the tubes, or with a grid connection on the side and the anode at the top.

A few weeks before the experiments with these tubes began the laboratory had built a serviceable Huth–Kühn circuit with two Telefunken RS31 triodes. These tubes proved unsuitable for waves shorter than 3 m because of the long, thin, grid leads entering through the tube base. Röhrig undertook the rebuilding of the RS31 to replace the monstrous RS207. The arrangement brought the grid lead out the side, and the anode was mounted so that dielectric losses were reduced and higher voltages permitted. In order to increase emission the heater power was doubled. In this construction one sees clearly the characteristics of the TS4 that GEMA was to create later. With these tubes 2 m wavelength pulses at 10 kW were generated.

Kühnhold was not favourably impressed that GEMA was working on transmitters for substantially longer wavelengths than had been used with the magnetrons. He favoured GEMA concentrating on designing decimetre-wave tubes in order to obtain more stable transmitter operation, and he was not prepared to make concessions for longer wavelengths with the resultant larger antennas. He insisted that centimetre waves were needed in order to produce a narrow beam with an antenna suitably small for practical use. He had derived from the experiments on 50 cm the characteristics required for transmitters and receivers working on 10 cm, and it appeared to him only a question of time before GEMA's experience with magnetrons would lead them to adequate powers below 15 cm. There was still a problem with the receiver, as the use of a super-heterodyne was essential in order to secure a marked increase in gain and for this a suitable mixer was needed.

In January 1936 Kühnhold gave a lecture to a selected group of naval officers about the state of development of sound and radio location. He reviewed the work then in progress, described the current achievements with location techniques, and explained how wavelength constrained design. He made a special effort to show the important value of very short wavelengths. His lecture left such a strong impression that he was allowed to continue his work on centimetre waves within NVA and in parallel to GEMA, which was by then loaded down with developmental responsibilities.

It is regrettable that the NVA, despite many microwave experiments, was unable to produce useable results. That Kühnhold, his co-workers and especially Dr Röhrl accepted the challenge of 10 and 15 cm electronics emphasizes the importance they gave it. They were supported by Pintsch and even by GEMA as best they could. Among scientists there was at the time, however, an opinion circulating that wavelengths in the centimetre band were unsuitable for radio location. Theoretical physicists in particular thought that too small a fraction of such waves would be reflected back to the transmitter, owing to specular reflection from surfaces not perpendicular to the direction of propagation. There was also the widely known instability of magnetrons.

The magnetron failures, as well as the adverse opinions of others, may have led Kühnhold to decide to pursue this line of research with decimetre rather than centimetre waves, thereby circumventing the need for magnetrons and using other tubes instead. Kühnhold's failures with centimetre waves for radio location in 1936 contributed to the mistake of the negative judgement on the applicability of this wavelength, an error not corrected until 1943 when a 9 cm radar set was extracted from the wreck of an English bomber near Rotterdam.

CHAPTER 9

A RADAR SUCCESS WITH AIRCRAFT

In the fall of 1935 GEMA installed the second radar of a series of three sets, on which von Willisen tried out an improved lobe-switching modification. Instead of using a separate receiver for each antenna, the two antennas were connected by a switch driven by a synchronous motor to a single receiver, the outputs being switched in synchronization to the vertical deflection plates of two CRTs in the indicator. The two screens were observed and compared stereoscopically simultaneously. A single receiver did not lead to an improvement in the bearing accuracy, but operation was simplified, because only one receiver had to be controlled, and any creeping variation of frequency and gain affected both signals equally and did not affect comparison on the screens.

Early in 1936 von Willisen tested a receiver with a 200 mm screen on which he observed an echo at a definite range. By adding an adjustable auxiliary pulse to the time base he was able to display a chosen echo in a magnified form. The auxiliary pulse was adjusted by the Messkette, and one could determine the range from its scale. The bearing could be read on a galvanometer with a mid-scale zero that was deflected by the sum of the receiver outputs for the two antennas, one of which was inverted. Rotating the antenna assembly produced galvanometer deflections indicating whether its axis was to the left or right of the target.

During these tests members of the NVA staff often took part, and Kühnhold was there from time to time. In addition members of various sections of Naval Ordnance began appearing in 1936 to look things over. Conspicuous and especially noted during the testing of lobe switching using the galvanometer indicator was the reduced interest of these personnel in any innovation that made the operation of the set more complicated. In the interest of simpler operation, demands on the accuracy of the radar were greatly relaxed. Through a policy speech in February 1936 at GEMA, Naval Ordnance established that, next to the vital completion of the S-equipment, the extension in range by the DeTe-set to 20 km was equally important but with direction determined by maximization of signal, not lobe switching, which was not desired. In addition to the Messkette, range was to be determined directly from the indicator screen, which was to be given a calibrated scale, which would simplify operation in case one did wish to rely on the Messkette. A new oscillator tube must replace the magnetron.

At the end of 1936 Kühnhold visited GEMA in Berlin in order to learn how the development of decimetre-wave triodes as replacements for the magnetrons was progressing. By that time the tube laboratory could show the first model of a transmitter tube whose electrodes were mounted on 1.5 mm diameter molybdenum rods, set vacuum tight in a hard-glass plate. Through this construction it became possible to make the external connections short and reduce loss. A small bell jar covered the electrode system and was fused to the plate. The first tubes worked in a push–pull Huth–Kühn circuit on 90 cm with an agreeable efficiency. Kühnhold was pleased with the result, even though he found the longer wavelength compared with the magnetron not to his liking. He still insisted that the goal of this research should be the creation of tubes and oscillators that were to be pushed to much shorter wavelengths.

He was really upset when shown a 15 kW pulse transmitter using RS207s for 2.5 m. He was not only annoyed by the large chassis with its 50 cm high output tubes but also by the imagined size of the equally magnified antenna, which was not suitable for shipborne service, and he declined to support metre-wave research completely, regardless of the temptingly high powers. Naval Command had accepted the large antenna size for the moment as a consequence of design constraints imposed by decimetre waves, but agreement for one with an area of possibly more than 20 square metres, implied by these wavelengths, was simply not to be considered. Kühnhold fell into a loud argument with von Willisen during which he threatened to favour Pintsch over GEMA, if they did not support him in his endeavour to obtain shorter wavelengths.

Von Willisen did not allow Kühnhold's anger to shock him, and stressed tests of radar on the 2 m band. Inasmuch as GEMA had no research grounds of its own, a little courage was needed to set up the new set on the tower at Pelzerhaken. This location also had the advantage of allowing comparison measurements with the decimetre sets there. In order not to irritate Kühnhold and his close associates at the NVA, von Willisen equipped the transmitter with rebuilt RS31s that gave it 7 kW on 1.8 m, and in order to prevent the NVA view of the tower from being disfigured by giant antenna walls, von Willisen had Yagi antennas installed just for this experiment, which with their seven thin elements mounted on a dainty mast were not particularly offensive. These antennas were not mounted on the turntable but attached to the tower, vertically polarized so as not to stand out. There was another reason for making them inconspicuous: the large size of a dipole array for 1.8 m wavelengths would have certainly excited the curiosity of unauthorized persons, which could lead to complications regarding secrecy.

The original installation called for the beam direction to be towards the woods at Scharbeutz, the location of targets that had by then become favourites. During the adjustment von Willisen realized that there were more ground returns than

48

in the earlier experiments with 52 cm. Because of tube failures, experiments with *Welle* were postponed until the beginning of March 1936. By then the antennas were directed toward the Mecklenburg coast near Klütz. *Welle* was sent into the beam several times, but an echo could scarcely be seen over the noise on the time base. Even large ships in the Lübecker Bucht were only weakly displayed on the screen. Sheer desperation soon became obvious. The equipment was checked and rechecked, but nothing was wrong, and no improvements were accomplished: there was nothing to be seen, and hardly even anything from large ships.

Because the blame that pointed straight at him plagued von Willisen, he quickly sought support from the laboratory in Berlin. Could not an overshooting direct signal be the cause of the failure, or was perhaps the input protection of the provisionally rebuilt receiver inadequate? As von Willisen, Rindfleisch and some co-workers from the NVA—fortunately, Kühnhold was not among them—morosely looked out onto the Baltic Sea and onto the indicator, they heard from Travemünde the sound of an aircraft that leisurely flew past at great range on its way towards the open bay. What was that? At 8 km on the calibrated time base of their screen an echo appeared that became greater and greater. Indeed, the pulse grew in size to the top of the screen, an echo the likes of which they had never seen. There seemed no doubt: it was the echo of the aircraft, which happened to pass through their beam at that moment, that was producing the huge signal. The indicated range looked right. Just as the echo had grown as the plane had entered the beam, now it shrank as it exited toward the sea. The departure of the machine was clearly shown in the vanishing of the echo signal.

There now began extraordinary activity on and below the GEMA tower. What had happened? How could it be that ships showed almost no reflection in contrast to an aircraft? All their discussions and conjectures failed to solve this riddle. It would have to be solved experimentally.

The next day a Junkers W-34 was chartered and ordered to fly parallel courses to that taken by the plane the day before, and the results were the same. As the machine flew across the beam of the radar at ever-increasing ranges, it could be followed as far out as 15 km, the distance to the opposite coast. Von Willisen then had the W-34 fly on a southeasterly course toward Grevesmühlen in Mecklenburg, the direction in which the radiation pattern was directed, and it was tracked with accuracy for 28 km; thereafter its echo dived up and down into the noise on the time base, and was finally observed at 30 km.

The first clue about the cause of the failure to observe ships was extracted from the data taken while tracking the plane along the radiation pattern. The higher the aircraft flew as it approached the tower from behind, the later its echo appeared beyond the zero of the time base on the indicator screen. These observations made it clear that the much-nearer ships, which had been hardly observable, had simply 'flown under' the vertically polarized radiation. Inclining the antennas slightly downward corrected this; now the passing ships were clearly seen on the radar screen.

After the electronics had been given another careful adjustment, the antennas were re-positioned on the tower so as to point over the bay in the direction of Schwerin. Now an aircraft was tracked to the range limit of the indicator, 'only' 40 km. The size of the signal at the end of the scale allowed the estimate that the then-current pulse repetition frequency of 2000 Hz formed the limit.

This astonishing improvement was so much beyond previous hopes that it swept away all the old NVA prejudices about metre waves. They were very interested in these results, and even Kühnhold came over with other naval officers to examine the 'long-wave' radar. Notwithstanding the clear need for decimetre radar aboard ship, there now appeared to the strategically minded officers the plan to supplement the naval tactical decimetre wave equipment with 2 m equipment for the long-range detection of aircraft. They presented the concept of radar for land stations to protect naval bases from air attack and for which the large antennas presented no problem.

Joined to the experiments for range were others to test the directional capabilities of metre waves. The two Yagis were mounted on a mast fixed to the turntable on the tower so as to allow directional adjustment. In mid-March 1936 a new set stood ready. The pulse width was reduced to 1 μs, and it had the first model of the GEMA triode TS4, designed for an 8 kV anode voltage. A provisional indicator was set for a maximum range of 75 km. The time base was sinusoidal, so measurements at the beginning and end of it were highly distorted, and even coarse measurements could be made only in the region from 15–60 km. A Messkette for these ranges was not ready.

At the end of March the W-34 stood ready to be a flying target again. In a series of experiments with a variety of antenna arrangements, the plane could always be located within the range limits imposed by the time base. Even at the maximum range, direction could still be determined by maximizing the signal. It was then decided that the range of the present Messketten should be doubled to 150 km.

With every increase in output power came more ground returns, and sometimes these echoes were so strong that tracking the target became extremely difficult. Von Willisen ascertained that, by tipping the antennas upward, he could find a position that reduced the disturbance of the ground echoes substantially. The experiments with Yagi antennas were ended, and dipole arrays, their 'Tannenbaum', were constructed as fast as possible.

The happy course of the experiments with long-wave radar had disposed of the tense relationship that had arisen between Kühnhold and von Willisen. His perseverance had unexpectedly laid the foundation stone for a form of radar that was superior in use against aircraft. It was characteristic of the lack of co-operation between different arms of the Wehrmacht in the 1930s that the Navy made a decision on a matter that affected air defence, which was the responsibility of the Luftwaffe. Only later did the Luftwaffe learn and take interest in the GEMA long-wave radar and thereby initiate the development of the legendary 'Freya' series. By 1945 from this basic 'Freya' arose all of the radar equipment for early warning, equipment that allowed enemy aircraft to be picked up as far away as 400 km, which meant about an hour's flying time for bombers.

In 1936 Erbslöh already had the idea of using a continually rotating antenna to provide a panoramic presentation. He was later able to develop this idea, as the 'Freya' evolved on the 2 m band.

CHAPTER 10

THINGS MOVE FORWARD

Beginning in 1936, things took a definite upward turn in matters of sound and radio location.

From the presentation of the preceding September and in the subsequent discussions, Naval Ordnance recognized they had found an energetic firm for changing Kühnhold's ideas into realities. Naval Command and the Chief of the Fleet had recognized the technical advances initiated simultaneously in sound and radio location and had instructed Naval Ordnance to give them high priority. Because all were convinced that GEMA's decimetre radar was unique and because of the favourable impression gained of the company's capabilities, the Navy sought to work with them on both location techniques. At that time, however, the Navy and its subordinate structure had an eccentric way of thinking that made them uneasy in dealing with GEMA.

Confirming the contracts, which GEMA was to carry out for the Navy, was regulated by extensive rules that governed collaborative work, which needed a long time for preparation, and that consequently kept them from going into effect until 1 January 1937. These agreements made GEMA one of many independent, voluntary partners of the Wehrmacht Fiscal Office. The favour accrued as the developing company of the two location systems brought with it certain restrictions on their business freedom for Erbslöh and von Willisen, the consequence of defence matters and their attendant secrecy. In concluding a contract that pledged the Navy to place a large number of orders, thereby requiring GEMA to employ up to 120 persons, GEMA needed guarantees of financial safeguards. The contract was to be in effect for five years and could not be cancelled until 31 December 1941. This provided them with a tangible view of the future for their company, and allowed them to plan on a long-term basis.

Especially after agreeing to work together, but even before, various naval bureaux distributed orders to GEMA. These concerned primarily the fields of communication and navigation. Naval ships and land stations already had GEMA apparatus for decimetre radio- and affiliated wire-telephone connections. In use was a decimetre radio telephone with speech encoders for secret communications between warships. Under contract for TVA, GEMA developed and fabricated special measurement, transmission and switching units. A special device was the difference-gyrocompass 'daughter', which showed deviations between gyros and which was duplicated on warships for safety. When the deviation exceeded a given amount it triggered an alarm that indicated one or both gyros were not in order.

Another contract concerned the final development of sonar, including delivery of prototypes ready for mounting aboard ship. According to the NVA booklet of requirements, GEMA was to deliver a sound-pulse generator with a frequency of 10 or 15 kHz and a power of at least 2.5 kW as well as a console with twin amplifiers in a sum–difference circuit and an indicator for direction and range. An effective range of 4–5 km was demanded. Development and delivery of the sound transducer and of the lifting and swinging apparatus was the responsibility of the firms Atlas of Bremen and Elac of Kiel. GEMA was the principal contractor and was responsible for the functioning of the total sonar apparatus. They were also responsible for the duplexing apparatus, which was necessary for the operation of the transducers. This passed or inhibited signals depending on the degree of magnetization of a laminated nickel core and is referred to a 'polarization rectifier'.

The other contract concerned the final development of a DeTe-Gerät for tactical use at sea, later simply named 'Seetakt'. For it the NVA had not written specific requirements to be met because continual improvements in range and accuracy could be expected, but Naval Ordnance did require the delivery of single DeTe units to be mounted aboard ships for testing. A range of 20 or perhaps 40 km was stipulated as a basis for these sets, and in order to simplify operation, the direction was to be obtained through maximizing the signal, not lobe switching. The Messketten should be uniform with a maximum setting of 40 km with steps of 0.1 and 1 km. The transmitter was to have the shortest possible wavelength. The shape and size of the antenna was not specified. Models should be constructed and tried out at the shipyard.

After the successful experiments with metre waves were made, Naval Ordnance expanded the development contract for the DeTe-Gerät with an appendix for an air-warning device intended for a land station. Seetakt was designated DeTe-I and the air-warning set designated DeTe-II, also called 'Flum' (Flugmeldung, air reporting).

Construction of all these devices had to follow the standards pertinent to equipment intended for warships. These were so all-encompassing that the relevant bulletins filled the boot of von Willisen's car. The GEMA staff was young and inexperienced in such specifications. This seemed a disadvantage for them in comparison with companies having had long experience with the Navy, but it proved rather to be an advantage in not forcing them to follow well-worn paths. With consultative support from NVA and other naval groups, GEMA produced constructions that were to become typical of them and not copies of others. It lay in the nature of things that to duplicate units from Seetakt for the air-warning equipment was construction nonsense, even laughable. GEMA made 'battleships' for the Navy. They had neither the need nor even the idea of making lightweight sets for aircraft. The distinctive construction of their equipment, which corresponded to Navy stipulations, inclined them toward being heavyweights and earned a reputation of great reliability.

More than 50 workers involved in the development and fabrication of the prototype and production sets were insufficient to meet the 1938 delivery goals for sound- and radio-location equipment, but the labour market allowed the employment of about 30 more qualified workers. The centre of gravity for the increases were Schultes' high-frequency and Brandt's low-frequency laboratories. Both men were excellent scientists who attracted thoroughly qualified technicians to aid them. Also successful was Röhrig's laboratory, where his creativity was of great use in inventing new tubes. During the time of co-operation with Lorenz he had supplemented his skills with the vacuum-tube craftsman Löpp. When the Navy ordered the suspension of co-operation with Lorenz, it proved in fact advantageous for GEMA in the development of tubes, because Röhrig followed a different path toward decimetre-transmitter triodes than Löpp, who had let the American acorn tubes be his guide. Röhrig designed a system for mounting the electrodes directly to connecting pins that were fused into a hard-glass plate. He examined diverse compromises to procure minimum electrode separations acceptable for the high voltages required. In contrast to Löpp he used directly heated thoriated-tungsten cathodes in his decimetre triodes, as he had used in his magnetrons.

In the summer of 1936 the first transmitter with the newly developed TS1 decimetre triodes was ready. It was anode modulated and delivered on 60 cm a high-frequency power of 700 W. This transmitter was built into a Seetakt set in Pelzerhaken and adapted to the antenna on hand. *Welle* was picked up at the first attempt at 7 km. As the Navy had no interest in lobe switching both receiver antennas were connected to form a reception mirror of eight vertical dipoles. This change, plus using the improved TS1 tube in the transmitter, increased the maximum range for detecting the *Welle* to 15 km. The tubes could now be modulated with 6 kV and delivered a satisfactory high-frequency power of 1 kW and more. Large steamers that passed were tracked to the end of the 20 km scale for which the set was calibrated. The set remained in this condition for some weeks at Pelzerhaken. The set aboard the *Welle* continued to use magnetrons because the transmitter with triodes was too unreliable.

Through the increase in the number of employees and through the creation of new departments, the GEMA Corporation became crowded on Potsdamer Strasse. A search for suitably spacious quarters in which the company could expand led to a three-storey factory building of the Accumulatoren-Fabrik AG (Storage Battery Factory, Inc.) in Berlin-Oberschöneweide, Gauss Strasse 2, and after some renovation GEMA moved there in 1 August 1936. Now in addition to the fulfilment of small orders, they could begin production of sound- and radio-location equipment in a big way. The construction department was expanded to include a model shop. To provide a secure housing for the electronics, for reasons of stability and because it was a cost-effective fabrication, GEMA began very early on using cast-aluminium cabinets of their own manufacture. They decided on the principle of dividing the various sub-units of a set into modules that could be self-connecting by being inserted into cabinets for which the plug contacts were made initially from parts obtainable on the open market. Later GEMA developed their own distinctive connection methods. Robustness and reliability, as well as simple maintenance and operation, were aspects on which von Willisen had a strong influence. During the past few years he had assimilated great experience in active radio location under all conditions, which he continually imparted to the developers and constructors in his company.

In the course of 1936 the radar was improved and made more reliable through additional experimentation. The 20 km effective range of Seetakt was confirmed. By fall the transmitter power with the improved TS1 tubes rose to 2 kW, and one often tracked targets to the scale limits of the set. In an experiment four standard dipole-array antennas were connected together to form a unit of sixteen equal-phased dipoles, and the effective range rose to 25 km for a medium-sized freighter. The arbitrary scale for Seetakt was raised to 40 km.

The development of the TS4 transmitter tube was well advanced. It already provided 8 kW on 1.8 m and reached 10 kW for 2.4 m. No further metre-wave experiments were undertaken with the Yagis because the results were so convincing that the NVA ordered additional experiments on three of these sets without hesitation. Although an antenna area of 25 square metres seemed enormous when compared with that of Seetakt, they planned a design with six dipoles each for transmitter and receiver. For reasons of uniformity the pulse repetition rate of 2000 Hz was retained for DeTe-I and DeTe-II. Changes in design for the air-warning set from that of Seetakt were accepted, which extended the range to 75 km.

The new plans foresaw that sufficient common components for DeTe-I and II were to be ready by early 1937 for ten units. Three transmitters and receivers for DeTe-I had to be ready then as prototypes, which were to be installed aboard ships for trials at sea. The antennas for sea duty were to be formed of steel-tube construction with metal screens as reflectors, in front of which an array of two groups of ten dipoles were each arranged for the transmitter and receiver. A new receiver was designed for the air-warning set. Both DeTe receivers used the double-frequency IF of 15 and 7 MHz, which became a standard. With this unit, receivers with decimetre-wave or new metre-wave inputs were followed by the same circuit units. For the decimetre input they retained the proven Q-multiplying mixer and oscillator using acorn triodes, which were by now being manufactured in Europe. The metre-wave input employed a preamplifier and a pentode mixer; the local oscillator was a triode; all tubes were acorns.

In addition to the rapid progress in different aspects of radar, GEMA could demonstrate in 1936 a good solution for the final form of sonar. NVA had sought for a long time with the help of GEMA a usable solution to the problem of instruments that would show together underwater range and, if possible, direction using the sum–difference technique. Fathometers had used lamps that moved with constant speed behind a ground-glass scale starting at scale zero on emission of the pulse and being lit when the echo was received. The scale was calibrated for depth, determined by the speed of sound in water. The lamp was generally a neon bulb, which could be lit by the receiver output, giving the method the name 'red light method'. Another somewhat similar procedure was sometimes used: the needle of a galvanometer would be released from its zero position by the outgoing pulse and began to move at constant speed; the returning echo stopped the needle, thereby allowing the distance to be read.

These methods were in principle usable for Kühnhold's horizontal ranging. Direction was determined by the phase comparison of the echo signals from the sum–difference amplifiers of the two transducers on their rotatable base. This required adding meters for direction to the one for range, and the NVA had attempted to develop a combined indicating device with help from GEMA. They used two fast galvanometers with mirrors reflecting light beams that were mounted behind a ground-glass scale that moved with constant speed. The sum–difference amplifier output signals would drive the galvanometers, deflecting the light beam onto the scale. For exact direction and range determination the sum signal was decisive. The transmitter pulse released the sled on which the ground-glass scale was mounted and a motor moved it slowly forward and then quickly returned it for the next cycle. The light from the sum amplifier indicated the range. Direction required a number of pulses and was obtained by maximizing the sum signal and minimizing the difference signal. Direction could not be determined with single pulses.

GEMA had, at the time, made a number of experiments with their various location devices using a CRT to investigate its usefulness for these purposes and, if satisfied, to prove its value, but they continually encountered naval rejection of the use of a CRT. **At the time of the first CW radio-location experiments with the audio-modulated transmitter they attempted to determine range from the phase difference between the modulation voltage of the transmitter and that of the reflected signal received. The transmitter modulation was connected to one pair of CRT plates, and the de-modulated receiver signal was connected to the other pair. This generated on the CRT screen complicated figures, the so-called 'Lissajous patterns'. For the same frequency of the voltages on the two pairs of CRT plates the figures stood still and changed their shape in accordance with amplitude and phase difference. Within a phase shift of 0 to 360° the figure changed in the following manner: from 0 to 90° a straight line running from lower left to upper right broadened through an ellipse into a circle (if the signal amplitudes were equal); from 90° to 180° it shrank through an ellipse into a straight line running from lower right to upper left; from 180° to 270° the lines passed through an ellipse to** a circle, and finally from 270° to 360° in the same way back to a straight line.

GEMA tried to use this procedure for coarse radar range determination by an estimate from the Lissajous figure; it could not be done accurately unless the transmitter signal was completely blocked from the receiver. They improved the method by inserting a phase-shifting circuit, calibrated in range, between the receiver and the CRT. Adjustment of the phase shifter to produce a minimum allowed the range to be read.

The Lissajous method essentially formed GEMA's concept for the indicator console for the sonar. The pattern was formed by the voltages of the two amplifier channels. The voltage of the difference amplifier was applied to the horizontal plates, and that of the sum amplifier to the vertical. If the transducer pair was properly directed, the CRT trace formed a vertical line. If the transducers were directed off axis, this vertical line inclined to the left or right. Adjustment by the operator was quite simple.

The second, novel part of the concept, which acquired the name 'Kippspiegel-kino' (swept-mirror movie) in naval jargon, grew from an idea of Erbslöh and his capable plant manager Golde.

In spring 1936 GEMA employed the firm Zeiss in Jena for furnishing a stereoscopic optical system for lobe switching. In using the system to compare the two signal strengths they had found this method useful and simple to employ. During the course of discussions that Erbslöh and Golde carried out with specialists at Zeiss about this system, a solution that employed a half-silvered mirror was proposed. Such mirrors allowed the comparison through a single eyepiece of the two CRT signals, one trace being observed on top of the other.

This proposal for a stereoscopic view carried Erbslöh and Golde to the solution of the problem of how one could put both the adjustment of direction and determination of range in the operator's field of observation. They examined their ideas in the laboratory and constructed a device that they were able to present to the Navy in December 1936 as an indicator for the sonar.

The figure that appeared on the CRT from the sum–difference amplifiers was viewed on a half-silvered mirror. A centrifugally controlled motor moved this mirror continuously over a cam, slowly forward and rapidly back. The mirror reflected an illuminated scale that was part of the arc around the mirror's axis of rotation and which corresponded to the apparent picture of the mid-point of the CRT screen as seen by the operator. Through the half-silvered mirror a given point on the CRT display was seen by the observer to lie on the scale. The rotation of the mirror was such that the mid-point of the CRT display moved steadily from zero to the end of the scale with constant angular speed. On the axis of rotation of the mirror was an electrical contact that keyed the transmitter on at the zero point of the scale to the zero point. Owing to the relatively slow propagation of sound in water, on average about 1740 m/s, the use of the mechanical mirror to draw the time base proceeded without a problem. For a target at 5 km the mirror had to rotate for 6.8 s to provide the necessary scale. **This was physiologically an agreeable period for the observer as it did not strain him. With exact adjustment of the transducer base the echo was momentarily represented with a vertical standing line on the screen that was matched to a point on the scale that gave the range. If the target lay to the left or right of the microphone-base's direction, there was an inclination of the trace that was easily recognizable. After a little practice the operator of the equipment could adjust the direction by viewing the appearance of the figure.**

This solution found acceptance by the Navy, and Kühnhold was greatly pleased that his fundamental work of many years had quickly found a crowning conclusion.

1936 ended with the designs of the two DeTe devices concluded, and the report on the sonar with its indicator could now be finished.

During planning meetings Erbslöh and von Willisen learned that the Navy was striving for the rapid installation and testing of this new apparatus on their ships. All large new naval construction was to be designed with space set aside, so that radar and sonar could be installed later without problems. The new devices were also planned for installation on smaller vessels. For the moment one could assume that the dimensions of radar antennas might be substantially reduced in the future, if centimetre waves became practical, but the known failures along these lines forced the Navy to insist on ending the tests and on establishing the details of the Seetakt antennas. At the same time they insisted that the location on the various ships for placing these devices should be determined.

CHAPTER 11

YEAR OF DECISIONS

A series of events decisively influenced GEMA in 1937, affecting in many ways both the form and the deployment of the two location systems. Radar took a direction that was a compromise between simplicity of operation and high performance.

With the potential by then available from more than a hundred employees, GEMA concerned itself with improving the range and the accuracy of both the radar and sonar for the Navy. Von Willisen applied himself not only as a representative of his company but also attempted through his sense of duty to explain the current state of the radar's development to all interested parties in the Navy and to show it in operation. In this he so conducted himself that, even as a civilian, he entered into discussions that concerned the deployment and use of the new device. He frequently encountered strongly expressed yet contradictory opinions.

Fleet Command had their own ideas. They had wide-ranging thoughts and put considerable importance on radar for naval tactics. By providing a way to see in the dark, radar forced all previous classical concepts of naval warfare to be rethought. For tactics on a large scale, ships had to be detected all the way to the horizon, a requirement that did not require great accuracy in bearing. Determination of range with the Messkette to 100 m was satisfactory.

But sections of Naval Ordnance held opinions that were not in agreement with those of Fleet Command. They wanted above all to see radar made into a method of fire control for various weapons. They had recognized in the demonstrations of the accurate determination of direction and range that it was within the domain of the possible not only to equal the accuracy of optical aiming systems but even to exceed it. The advantage of being able to aim independently of visibility spoke loudly for radar. Given these requirements for a universal method of fire control for artillery and torpedoes, the range achieved so far of 20–40 km was completely satisfactory. Naval Ordnance pressed the requirements of a directional accuracy of 0.2° and a range accuracy of 50 m. Attaining these accuracies, however, would give rise to long delays before their realization.

In extensive correspondence between Naval Ordnance and GEMA in autumn 1935 these concerns animated the discussions. By early 1937 it was recognized that a radar employing two antennas for lobe switching could not be ready in a model suitable for duty at sea within the year. The decision about an applicable technique was therefore delayed until GEMA could complete various experiments. Thus high directional accuracy was not considered for the time being.

At the time when the research boat *Welle* was being outfitted with a new, improved radar, Naval Command insisted that a torpedo school boat, an artillery school ship and one large naval unit be similarly equipped. Also the NVA and TVA should each receive a new Seetakt set in order that they might resolve the assorted differences of opinion concerning the use of radar in the Navy.

GEMA did not allow the conflicting opinions of the Navy to distract them. They accepted improvement in accuracy as an absolute necessity, and certainly had no need to worry about their principal contractor, Naval Ordnance, who approved and supported all the work intended to improve accuracy.

Since the introduction of the pulse technique in early 1935, GEMA had concerned itself with the details of time-delay measurements. In contrast to sonar, for which the travel time of a signal from transmitter to target to receiver was rather long, about 7 s for a reflection 5 km distant, the time delay for a radar target 20 km distant was 133 μs. For measurements of such short time periods only the CRT could be used. There was simply no other way, even if the Navy bristled at the use of these tubes aboard warships. Although Ferdinand Braun, for whom the CRT is called the Braun tube in much of the non-English-speaking world, invented the device in 1897 and showed how it could be used in recording very fast processes, it had found wide use only recently for television. The improved sharpness of focus and the distortion-free deflection had made it useful for electrical measurements. With CRTs that could be purchased from the electronics market, GEMA undertook experiments with various circuits for producing the time base. Especially important was the ability to read the range as accurately as possible from the position of the echo on the calibrated scale. At the time the largest CRT screen diameter was 200 mm, tubes with larger diameters becoming available only later. The accuracy of the time measurement was determined by the rise time of the signal. The distance on the screen between the zero position of the trace and the flank of the echo needed to be as long as possible, so the echo might be recognized and marked. In principle a CRT with horizontal time base for the whole range of delay times sufficed for an approximate measurement of distance, but for the requirements being demanded this resolution was inadequate. Experiments with circuits to extend the range of the time base were not successful.

With a new tube from the laboratory of Leybold & von Ardenne, GEMA tried out a new way of writing the time base. In this tube the cathode ray described a circle on the screen and allowed a representation in polar coordinates. The circular deflection was provided by two magnetic coils mounted perpendicularly to one another at the neck of the tube. With an *RC* phase-shifter the current in the two coils was shifted 90° relative to one another. This afforded a path length about three times longer than a straight line, as a sinusoidal voltage allowed an uninterrupted movement of the spot with constant angular speed. These advantages had as drawbacks the imperfect development of these first polar-coordinate tubes. With the early tubes and deflection circuits, it proved difficult to draw an exact circle on the screen that could be held accurately concentric with the circular range scale. To improve the design of such tubes themselves would have demanded too much. There remained the risk from the unresolved question of when, or even if, the Navy would allow the CRT to be used as a radar component, although they had accepted it in principle as an inertia-free indicator.

After weighing all the benefits and disadvantages GEMA decided to use a linear CRT display. **The time base would use sinusoidal voltage. The lateral deflection to the left and right of screen centre was made with an audio-frequency generator that provided the deflection amplifier with a voltage that drove the cathode ray well beyond the edge of the screen on both sides. The portion of the voltage that was observed on the screen was sufficiently linear for a range scale. A special circuit suppressed the brightness of the spot on its return trip. The horizontal deflection voltage was rigidly fixed to the phase, that is, the zero of the horizontal time base, which could be extended greatly by expanding the amplitude of the sinusoidal voltage, always found itself at the centre of the screen. There was a fixed and an adjustable phase-shift circuit between this audio generator and the pulse modulator so that the pulse could be given a phase lag of $-90°$ to 0. The zero point could thus be placed at the left edge as well as the middle of the screen. For coarse measurements of range the entire length could be used. When accurate data taken with the Messkette were required, its zero was set with the transmitter pulse at the marked CRT centre.**

The method of making accurate measurements consisted of using the Messkette to shift the phase of the audio signal being fed to the deflection amplifier, that phase shift corresponding to the time between the transmitted and received signals. Thus the echo pulse could be positioned in the middle of the screen by the Messkette. The transmitter pulse previously lay at this zero point when the Messkette was set at zero. This allowed the range to be measured by comparison, permitting higher accuracy than by reading the trace directly.

The simplest Messkette was the *RC* circuit made of resistors and capacitors, but a given delay was strongly dependent on frequency and the output voltage was attenuated and its amplitude frequency dependent. Such a circuit was rejected by GEMA from the beginning. They followed the advice of Professor Salinger of the Heinrich Hertz Institute and used a 'Wagner transmission line', consisting entirely of coils and capacitors. This circuit was indeed extravagant but was also nearly independent of frequency and the output voltage was the same for all delays, large or small. The delay

times and the steps demanded could be calculated and spread out as desired.

A special advantage of this GEMA method was that the resolution in range was limited only by the fine graduations of the Messkette and was independent of range. The accuracy of the Messkette over long periods of time and with variations of climatic conditions, and the accuracy with which the transmitter and echo pulses were marked, determined the accuracy of the radar for range. GEMA would not have pushed the development of the Messkette had Naval Ordnance rejected the use of all but the smallest CRT.

By the end of 1936 the Kabelwerk Oberspree division of AEG completed the toroidal coils for the Messketten. Having this work done outside hindered the work of the low-frequency laboratory and cost additional time. At the beginning of 1937 Brandt procured a winding machine for toroidal coils and improved the production of the coils and the flexibility of the circuit design. The firm of Jahre delivered mica capacitors of the highest accuracy and time stability. The range steps of the first Messkette were in decades that were switched individually, but the low-frequency laboratory invented, together with the construction section, a new switch arrangement that relied on the Geneva mechanism, well known in moving-picture technology. In this arrangement, switching the range simultaneously set the data wheels. On the basis of the experience the company had had with the fabrication of cast-aluminium construction, they enclosed the components of the Messkette—the coils, capacitors and resistors—in a cast-aluminium housing rather than in a chassis box formed from folded sheet metal. On the front of this frame was located the drive coupling and the data wheels.

Not to be forgotten in describing the elements that led to the accuracy of the Messkette was the central audio-frequency generator, which provided the timing for the modulator pulse and for the deflection of the time base. Not only constant frequency but also a constant output voltage with good sinusoidal form was required of it, as harmonics and amplitude variation in the deflection led to errors, especially in range measurements read directly from the screen. The audio generator worked from a stabilized special type of oscillator that used an AC2 tube and a temperature-compensated resonant circuit made up of toroidal coils. An additional AC2 served to isolate it from the tuned output amplifier, which also used toroidal coils.

GEMA experimented with indicators outfitted with picture tubes made by Radio-AG Loewe. This company had brought out a tube with a 200 mm diameter screen that had two cathode-ray beams. GEMA intended to use the two systems for coarse and fine ranging. The tube stood vertically and was observed by means of a mirror and magnifying optics. A range scale was marked on the screen at the time base. In the middle of the screen was a fine mark for use in zeroing the Messkette. A potentiometer served to control and adjust the deflection amplification in order to make it correspond to the calibrated scale. For calibration, the sinusoidal voltage was adjusted to fill the screen between two markers. After this calibration the amplitude was changed through the use of a resistive voltage divider, so that a fixed range was represented on the screen.

The receiver used was an older model in which the HF and IF were not built into separate mechanical units. It was so well screened inside that the high-gain pentode 4673 could be used. The bandwidth was 700 kHz.

The decimetre transmitter was decidedly improved by employing the new TS1 output triodes. Good business relations with the Hescho Company of Hermsdorf, Thüringen, which had delivered ceramic capacitors to GEMA, led to a collaboration in producing moulded, ceramic insulators. Brandt had Hescho fabricate ceramic housings in which he enclosed a precision mica capacitor for the Messkette. By means of copper rings that were brazed to the ceramic, he could solder the cover so as to protect the capacitor completely from its surroundings, assuring long-term stability. This gave Golde the idea for his so-called 'gebackenen Schwingkreis' (baked resonator). First he proposed the idea of the 'ultra-resonator', the complete generator with TS1 tubes in push–pull for the decimetre transmitter. The capacitor of the tank circuit was fixed to a copper mount that was brazed to a ceramic plate. On this plate were found the connecting sockets for the TS1. The final configuration of the ultra-resonator had two parallel ceramic plates held with spacing rods screwed onto the four corners; they served as the bases for the TS1 tubes and as the mounting for the other components. Between the plates were located the anode and grid rings that formed the tank coil of the Huth–Kühn circuit. The positions of the two triodes were arranged as mirror images, which required a special version of the TS1, the TS1A, that had the required mirror-image arrangement of its connecting pins relative to the TS1, so that the grid and anode of the two tubes faced one another in the assembly. With this unit, called the 'Ultrateil', the new transmitter worked very reliably with a power of 1.5 kW. (The reader will find reference to figure 12 useful.)

In spring 1937 GEMA set up new versions of DeTe-I and DeTe-II at Pelzerhaken. The DeTe-I was placed on the GEMA tower with rotatable antennas for transmitter and receiver for a trial of the arrays of ten full-wave vertical dipoles. The DeTe-II was set up at the base of the tower in a wooden shelter and had antennas for transmitter and receiver of six full-wave vertical dipoles each that were fixed in direction toward a reference point south of Rostock.

The maximum range of the DeTe-I for maritime targets was, depending on the size of the vessel, 10–25 km. The DeTe-II tracked a Junkers W-34 to its turning point at the River Warnow south of Rostock at a range of 80 km. The introduction of the Messkette allowed the range to be measured to 60 km. Because the antennas were fixed in direction, the aircraft was constrained to fly back and forth in the beam direction, but as far away as 50 km the echo was sufficiently strong to have allowed the direction to have been determined by lobe switching. DeTe-II was capable of observing ships that passed through its radiation pattern all the way to the opposite shore, with stronger echoes than DeTe-I on the tower, owing to the fivefold higher power and the greater receiver sensitivity.

Because GEMA had no testing ground of its own, the workers employed for their experiments had to make their way to Pelzerhaken, and since fall 1936 the numerous experiments resulted in a regular courier service being instituted between Berlin and Pelzerhaken. The Navy proved to be very choosy as to whom they permitted to participate in these experiments. Particularly exasperating was that GEMA was constrained to provide the Navy information about whatever work was going to be carried out, yet providing this information to other agencies, specifically other Wehrmacht agencies, was forbidden. This made it impossible for GEMA to undertake business contacts with the Luftwaffe, although they showed interest in radar.

In mid-1935 Admiral Raeder, Chief of Naval Command, was named Commander-in-Chief of the Navy. He was a zealous proponent of surface strategy, and since the demonstration of radar to him on 26 September 1935 he had been extraordinarily interested in the further development of this new technique, as it fitted in well with his strategic concepts. In the disputes between Fleet Command and Naval Ordnance he favoured equipping all large units of the fleet with one or more radars. When Raeder heard of GEMA's more recent work with enhanced ranges, he requested a demonstration of the Seetakt and Flum (DeTe-II) equipment.

At this demonstration GEMA showed the new antenna that allowed lobe switching. The experience with an antenna in which the principal beam direction was perpendicular to an array of dipoles excited in phase had led to a design in which the central connection to the ten receiver dipoles could be separated. This separation yielded two arrays of five dipoles, which, if seen from the division point, could be asymmetrically switched. In this arrangement the beam directions of the two arrays differed, owing to disparate travel times in their feed cables. This gave the same result previously obtained by physically altering the orientations of the arrays: the radiation patterns crossed one another. Having both arrays in the same plane allowed one to connect them together for maximizing the signal of all ten dipoles or to switch the lobes for accurate direction finding. With this solution GEMA had built an antenna suitable for both purposes. The light cruiser *Köln* served as a sea target for the demonstration of the Seetakt, which worked on 60 cm. The NVA research boat *Welle* was no longer available, as she had foundered in a storm.

By the end of 1936 GEMA had provisionally installed prototypes of their sonar equipment with the new indicator on U-26 and the coastal ship *Laboe*, which the NVA had chartered. In the first half of January 1937 trial voyages were conducted in the Lübecker Bucht during which the *Welle* took part as a communication relay. In the experiment room on board *Welle*, besides a functioning DeTe-I, there were accidentally the extra electronics for an additional DeTe-I. For the GEMA radar technician aboard *Welle*, who communicated with the participating vessels through the decimetre radio telephone, there was spare equipment and a radio direction finder.

On 15 January 1937 von Willisen went to Kiel to fetch the GEMA apparatus from *Welle*, as it was to be serviced and improved in the Berlin workshop and was then to be mounted on torpedo boats G-7 and G-11. The *Welle* had been ordered to Kiel on completion of the exercises with U-26 and the *Laboe* and was to receive a new radar and a new decimetre radio telephone.

For unexplained reasons *Welle* was posted missing for several days. A report was received that she had aided the motor ship *Duhnen*, which was in distress on account of a strong snow storm with fifteen Nazi Party members and a crew of eight on board. There had apparently been difficulties in persuading the Party members to abandon their ship and in the process *Welle* also came to grief and was lost with all hands. This accident had a macabre aspect. A few days after the sinking of the *Welle*, a Danish patrol boat found the corpse of the quartermaster and salvaged some wreckage on Falster Island, which they turned over to the Navy at Flensburg. Among the recovered wreckage was the CRT of the DeTe-I, its range scale intact. The circumstance that the tube had suffered no damage and still functioned after having been in the water and surrounded by drifting ice brought happiness to the champions of the CRT. There were less happy people who would have preferred that the perfidious CRT had lain smashed on the ocean floor, thereby leaving their theories intact.

Besides *Laboe* a small ferry boat was outfitted with rotating bases for transmitting and receiving underwater sound. After the current experiments and tests with the half-silvered-mirror indicators, which were used on both ships, had succeeded to the extent that all parties were satisfied with the solution, GEMA built a small series of ten sonars. They themselves participated in these test voyages only technically, as the set designs were so mature and dependable that their operation presented no problems. More than twenty sound generators had proven themselves in steady, continuous operation. The oscillator, driver and power amplifier using two RS285 tubes in push–pull were located in the upper level of one rack; in the lower level was the power supply. The sound pulse attained 2.5 kW and could be turned down to 25 W.

In the upper part of a second rack, intended for controlling and observing, was a CRT with its power supply and the parts of the half-silvered-mirror apparatus. In the lower half were the twin amplifiers with sum–difference circuits and their power supplies. Each of the twin amplifiers had four stages, providing a total voltage amplification of a million. In order to ensure a minimum of cross-talk, the two amplifiers had to be carefully decoupled and shielded. Brandt had given these amplifiers his utmost attention in the low-frequency laboratory and had derived a concept that allowed them to be produced in large numbers and operated without complication, which they did until the end of the war.

The voltages that the sound transducers could generate for weak echoes amounted to only 100 μV. The amplifier channels were resonance selective, but nevertheless the connecting cables had to be made symmetric as compensation against extraneous signals. Problems arose particularly when switching from transmission to reception. With mechanical switches and normal relays there was only a short time to avoid transition resistance during switching. The high power of the transmitter pulse and the corrosive atmosphere to which the switches were exposed required continual cleaning of the contacts

in order to allow the weak transducer signals to be connected without disturbance. There were vacuum switches, which had to be served electrically, but they were relatively large. GEMA therefore developed their own vacuum relay with switching contacts that were sealed into a small glass envelope. They were activated electromechanically from without through a small glass bellows.

In early summer 1937 Torpedo Inspection began the installation of DeTe-I sets in various vessels, among them the torpedo school ship G-10. This ship caused a stir when, at the end of the year, an uninformed photographer selected it for an appearance in *Taschenbuch der Kriegsflotten 1939* with the radar antenna easily seen. Only after publication was this blunder noted. The photograph had passed through the relevant naval authorities, who had no idea of the significance of the 'mattress' located forward of the foremast.

GEMA placed a special Seetakt set at the disposal of Artillery Inspection for experimentation. In co-operation with the Range Measuring School, which was responsible for naval rangefinders, experiments were undertaken to compare optical and radio methods of determining range aboard an artillery school ship and gain experience in using the new method of gun laying. Just for this purpose GEMA constructed an indicator unit with the new 200 mm dual-beam CRTs. With 1000 Hz pulse repetition rate, which corresponded to a range of 150 km, they could present 60 km on the screen with good linearity. The Messkette of the set also conformed to 60 km. New to the indicator was a supplementary deflection amplifier for the time base of the second beam. By extending the sinusoidal deflection voltage of the second beam, one gained a magnification of the time base of the echo pulse and allowed it to be seen next to the normal pulse. This allowed the exact placement of the echo on the zero mark of the tube, thereby allowing the Messkette to be adjusted more easily.

In addition to the light cruiser *Königsberg*, the NVA research boat *Strahl* was outfitted with DeTe-I and II for the September 1937 Wehrmacht (armed forces) manoeuvres. It was the first time that the large Flum antenna had been mounted aboard ship. Two additional DeTe-II sets were brought to the manoeuvres in Swinemünde and set up on the *Kückelsberg* and the *Brandberg*. The sets were built into a large van and a six-wheel communications truck. Transmitter and receiver antennas of the DeTe-II were mounted on rotatable bases furnished by the Zeiss Company and built especially for this purpose.

Next to the Seetakt set aboard the *Strahl*, which could easily cover distances out to 20 km, was set up a DeTe-II for air warning. The three DeTe-II sets used in the manoeuvres worked on 2.1 m and could have their pulse repetition frequency easily changed to 1000 Hz, in order not to interfere with one another. The indicators were set up for 80 km with Messketten for 60 km. With the Flum aboard *Strahl*, the aircraft types Ju-52 and He-111 were located, depending on their height, at 40–60 km and tracked to the 80 km limit of the indicators. The two sets on land functioned just as well.

Since the beginning of 1937 the development contracts and the increasing number of single orders, especially after the general naval contract went into force, so occupied GEMA's capacity that it had to expand yet again. When the number of employees passed 150 there was again a shortage of space, which had to be found somewhere besides at Gauss Strasse, if there was to be adequate working room. This problem presented itself to Erbslöh and the plant manager Golde. Von Willisen was to be found only presenting demonstrations and at consultations with the Navy.

After the radar demonstrations before the Commander-in-Chief of the Navy at the Wehrmacht manoeuvres, Naval Ordnance began to pressure GEMA by way of von Willisen, who was generally the person at hand, to increase their efforts in all areas. Through Commander (retired) von Simson, their friend and advisor, Erbslöh and von Willisen learned that the Navy intended to offer GEMA a loan for possible construction of a plant devoted entirely to development and production for the Navy.

Erbslöh and von Willisen until then had had no doubts that they did not want to give up their research and development nor to limit the production of models and special equipment. Their contract required them to disclose to the Navy at the beginning of 1937 what companies they would use as subcontractors. This led to the first, official talks about the intended expansion of GEMA into a production plant. The good experience that the Navy had had so far with GEMA, and the circumstance that the company had only two owners and did not belong to a concern, prompted Naval Ordnance to request that GEMA do their own fabrication as soon as possible. In addition to that, they should render help for a fee at naval bases and yards where they would provide assembly stations and workshops for installation and maintenance of their equipment.

After many discussions with the Navy and especially after long deliberations with trusted legal and financial advisors, Erbslöh and von Willisen decided to take the biggest step of their business life: they changed GEMA from a middle-sized development company into a substantial production company for sonar and radar equipment. They were able to negotiate an interest-free loan from the Reichs Fiscal Office for this expansion. The loan was restricted and might only be used for the erection of plants sufficient for the naval purpose. For insurance, a mortgage was drawn up against the owners and the plant and the creditors were relieved. The independence of GEMA and its owners remained complete. Erbslöh and von Willisen could hinder any attempt by the Navy to gain control of GEMA through acquisition of stock.

In September 1937 GEMA bought 36 000 square meters of factory grounds from the Linoleumwerken A.G. on Wendenschloss Strasse in Berlin-Köpenick. On this property, which was bounded at the rear by the Dahme, a tributary of the River Spree, stood buildings of the inactive linoleum factory with 145 000 cubic meters of space that were to be converted to their purpose by early 1938. On this property it was now possible for GEMA to expand significantly. All preparations for fabrication could now be made generously, according to need, and would not suffer from the shortage of space.

Based on the experience that he brought from his work with Görler, Golde began the preparation to expand GEMA. Serious interruption to the current development work could be avoided during the move, as the space on Potsdamer Strasse would not be given up and would continue to be used for a time. Model construction remained there and it took over the first production series of sonar generators in addition to decimetre radio telephones.

With the production of radar and sonar the Navy had to face the serious problem of training service personnel. Because work with this equipment was new and complex, practical training could only be made with the sets, and the teachers had to have extensive practice with the apparatus. Von Willisen had long established himself through presentations and orientations of operators, generally of naval radio personnel. He was a man of long experience and had gone through all phases of the development. For him there had been hardly a situation that he could not surmount. With his background, which he had collected from many trials and experiments, he had become the preferred consultant of the laboratory heads at GEMA as well as Kühnhold and his co-workers at the NVA. During the time of the testing phases that had elapsed, he was constantly among the naval personnel and was accepted as one of them.

Partly from caprice but also for plausible reasons, friends in the Navy persuaded von Willisen that he must join them as a means of improving collaboration. He entered the Naval Communications School at Flensburg on 1 December 1937. After that his working clothes were a naval uniform. It had not been his intention to have to wear these 'working clothes' to the bitter end in 1945.

CHAPTER 12

THE BEGINNING OF PRODUCTION

A new chapter began for GEMA in 1938 in which the character of the undertaking changed completely.

Erbslöh and von Willisen had attacked the tasks that Kühnhold and the NVA had presented them with, and through their rapid, supportive handling of these problems they had proved the correctness of his ideas. By rights they might see themselves as the midwives of sonar and radar, and in this they were certainly not conceited. It had enriched their lives to have been able to work with the physicist Dr Kühnhold and his co-workers in these completely new fields. Beyond this, they had true friends and partners and had provided support, competent advice and practical help with their skills as craftsmen. For constant backing they quickly surrounded themselves to the limits of their finances with accomplished co-workers.

As they entered deeper into the enterprise and as successes mounted, so did their problems, and the first GEMA generation, the one that still had close ties to the NVA, matured with these problems. Some worked in Berlin and some at Kiel or Pelzerhaken on the solution of common problems with Erbslöh and von Willisen supervising one side, Kühnhold the other. With the exception of occasional contrary opinions, there was a harmonious association between the two parties. The outcome of various trials was generally learned at Pelzerhaken, and every successful, but also every unsuccessful, result inspired work towards improvement. And just as Kühnhold inspired and animated his staff, so too did Erbslöh and von Willisen inspire theirs.

After its formation in January 1934 GEMA established their activities predominately based on part-time personnel in addition to Hollmann's collaboration, but began hiring permanent scientific and specialist workers. Thanks had to be given to Röhrig for their rapid, successful rise in the development and production of decimetre triodes and magnetrons. From the large supply of unemployed or underpaid scientists, engineers and technicians available at that time they were able to put together a staff of young, dynamic and skilled workers, which came not only from Berlin but from every corner of Germany. All had excellent chances for speedy advancement. By autumn 1937 GEMA had over 200 employees.

The company's founders had an engineering background, and though circumstances had forced them to become predominately entrepreneurs, they lived not over their co-workers but among them. A good working environment was found in their company, and people were stimulated. Problems, where they permitted, were solved through open dialogue. Patents and registered designs were encouraged and generously rewarded.

The first department heads, Dr Schultes for high frequency, Dr Brandt for low frequency, surrounded themselves with qualified co-workers with whom they worked with assurance and harmony. This made possible their rapid achievements in the new technological domains of sonar and radar. In the beginning they were often bolstered by personnel at NVA, but owing to distance this support eventually became an occasional consultation, generally when it involved something with regard to the Navy.

The secrecy imposed in 1935 resulted in GEMA being able to work without competition for a time in a field that was only imperfectly grasped by the larger firms. They were quickly able to secure the business through a number of patent applications, generally secret and many of decisive importance. Except for the time of their collaboration with Lorenz, who had obtained a good look into GEMA's work and patent structure, the results of their research remained hidden from other firms, whose opinion of this newcomer to electronics was not high, and who consequently made no effort to learn of their activities. Their possession of much long-standing electronics experience allowed them to dismiss young GEMA as a 'group of hobbyists'. Lorenz had noticed GEMA's strengths first and had sought to work with them. Telefunken initially considered the work in radio location at NVA and GEMA to be utopian but began their own radio-reflection experiments in 1935, which led to significant progress in 1937, when the first GEMA radar sets were being tried out at sea.

By 1937 at the latest it was known to more than just a restricted circle of persons that radio-reflection experiments were being carried out at Pelzerhaken and apparatus appropriate to their exploitation was being developed. Kühnhold's and GEMA's successes could no longer be kept so well hidden that outsiders got no wind of them. The Pintsch Company had exerted themselves to a high degree in attempting to improve their centimetre-wave equipment, so Kühnhold could attain serviceable radio-location results. They knew how close to his heart this approach lay; nevertheless they were unable to satisfy him. Kühnhold proclaimed over and over to Pintsch and GEMA the possibilities of microwave research, as this was his strong interest, but failures led him to discontinue this line of work with these two firms after 1936. The use of microwaves was by no means forgotten by him, and he continued investigations at NVA.

The Air Research Institute in Oberpfaffenhofen began in 1937 to study the reflection of centimetre waves and was able to observe a lake steamer on the Ammersee, using the Doppler effect to measure the vessel's speed. The breakthrough needed to use this technique, however, did not come.

GEMA was ahead of all the other firms in having a comprehensive research and development programme behind them, which had brought them much experience that the others had yet to collect. Even though tube technology had in the meantime made general progress and had enabled the generation and amplification of decimetre waves, there was still no recipe book that explained how to use them for location at sea or in the air. GEMA could not and dared not have an interest in giving up any of their knowledge. Their employer, the Navy, held knowledge of the new location technique secret even within their own ranks and did not share it with other arms of the Wehrmacht.

On deciding to manufacture the equipment that they had invented GEMA began the reorganization of the company. Until then their responsibilities had ended with the fabrication of operating models, and now they had to build prototypes suitable for manufacture in large numbers. They had to create a production line that was no longer limited only to experimental apparatus. In the future the economies of production and maintenance had to be given consideration too. After the successes that GEMA had had with the development of their sonar and radar, they constructed a factory for production with over 1600 workers in less than two years and built it from the ground up. They did not copy from other firms but went exclusively their own way.

On the land that GEMA had procured on Wendenschloss Strasse stood a collection of buildings built between 1885 and 1924, primarily large storehouses of massive brick construction and still in good condition. In the middle of the grounds between Wendenschloss Strasse and the Dahme were distributed a line of four large complexes with a fifth oriented at right angles to them. Some of these structures were intended as sheds. In mid-1938 Erbslöh proposed, according to his concept of the requirements for production, to undertake alterations that would have 10 000 square metres for this new sphere of activity. To the side of the central structures would stand buildings for a foundry, machine shop and for the domains of glass blowing, painting and galvanizing, as well as offices. In a detached building 1000 square metres of the first floor were reserved for large machine tools. In the floors directly above were housed the testing of units and material. The mounting of rotary bases and antennas was to take place under sheds outdoors, and a large hall suitable for assembly was planned for the left side of the property and completed in 1939. The administration was quartered in two small buildings along Wendenschloss Strasse and in buildings behind them wherein the design and drafting departments were to be found. A new four-storey administration building with 7000 square metres was planned and occupied in 1940. It was to be constructed at the left in an angle formed by the street and the assembly hall. Laboratories with associated research needs were placed at the end of the grounds and facing the Dahme in a building parallel to the stream.

An organization evolved in these surroundings whose strongest attribute was production. By mid-1938 production included the radio telephone, sonar and radar.

The radio-telephone FG consisted of a relay rack FGU, instrument rack FGI, receiver rack FGE, amplifier rack FGV, call rack FGR, modulator rack FGM, anti-side-tone rack FGG and power-supply rack FGN or battery rack FGB. Following the letters identifying the equipment came numbers with which the working frequency and the version were designated. Under the identifications FS and FE, GEMA produced antennas for transmitting and receiving for their radio equipment, which would satisfy the requirements of various sets. Group FS included the transmitter itself, in order to minimize cable losses. The wavelengths of the radio telephones lay between 73 and 80 cm.

The 15 kHz sonar consisted of a generator console HS 15 with oscillator stage HSV 15, driver stage HSS 15, output stage HSE 15, power supply HSN 220, remote control HF 15, control console HM 15 with half-silvered-mirror indicator HMK 15, cathode-ray tube with power supply HMB 220, twin amplifiers HMZ 15, power supply for twin amplifiers HMN 220, polarization rectifier and transmitter relay box HRS 15 and polarization rectifier and receiver relay box HRE 15.

The DeTe-I (Seetakt) and DeTe-II (Flum) radars consisted of a transmitter base with transmitter GSU, control unit for 2000 Hz GSS 2000, control unit for 1000 Hz GSS 1000, transmitter cooling unit GSL, operating unit for transmitter GB, phase shifter GBN, receiver base with receiver GEE, receiver power supply GEN, receiver amplifier GEV, indicator for 2000 Hz GEB 2000, indicator for 1000 Hz GEB 1000, Messkette GEK, audio oscillator for 2000 Hz GES 2000, audio oscillator for 1000 Hz GES 1000, antenna array GA with transmitter array GAS, receiver array GAE, antenna matching unit GAA, base for direction attachment ZS with direction apparatus ZSS, direction indicator ZSB, antenna lobe switch ZSU, transmitter power supply GB, high-voltage power supply GBH, power connection GBN and instrument panel GBB.

For special DeTe-II sets there was a separate audio oscillator unit designated ZB. It contained the oscillator, output amplifier and line connector ZBS, as well as power supply ZBN 1 and ZBN 2. With these oscillators several radars could be operated synchronously with the same modulator and time base.

Pressure from the Navy caused an acceleration of production at Wendenschloss Strasse, and deadlines pressed everywhere. Equipment was increasingly needed for radio-telephone connections between naval bases on North Sea islands but also for communications between ships. Testing the first production sonars went extraordinarily well, so that in this area there were extensive lists of ships in which sonar was to be installed. There were orders for 50 radars of various models for ship and land installation.

Although construction work was in full swing on the factory buildings, and the machinery as well as the tools, workbenches and auxiliary equipment were still on the way, the number of employees rose substantially in summer 1938 to pass 500. Only the fact that Erbslöh—von Willisen was now in the Navy—and his capable crew had reached a working environment that was almost like a family made it possible for the new organization to function, almost without friction. It proved very useful that he moved into a house that stood near the factory on Wendenschloss Strasse, for by living near the plant he could always be reached. Already by autumn 1938 an orderly growth of production could be credited to the new GEMA organization and the number of employees continued to increase as the end of the year approached.

A rotating base was constructed for DeTe-II. A round, wooden, operator's cabin of angle-iron construction was fixed on the gun carriage of an 88 mm anti-aircraft gun, in which the electronics of the set were protected from the weather. The transmitter and receiver antennas were affixed one over the other to two tubular-steel masts on the side opposite the cabin entrance. Directly behind the transmitter array was arranged a canvas sheet on a frame as protection against the weather where the transmitter was attached near the antenna feed. The radar was mounted to the gun carriage so that the electronics and the antennas were balanced relative to the rotary pivot, allowing the whole assembly to be easily rotated through 360° with a handwheel. From a horizontal angular scale one could read the direction. The rotatable base, which had been made in the GEMA shops, could be disassembled to the extent that the components could be loaded onto a truck. A special four-wheeled trailer, type 204, transported the gun carriage and was pulled behind the truck. Disassembly was necessary because the wooden walls of the first rotating bases did not yet give a small enough profile for passage along roads.

The antenna array had a construction with a wire-grid reflector that presented the smallest possible wind surface. The first arrays were fixed to hardwood frames. They looked like mattresses and were jokingly called that. For ship-borne construction metal tubes replaced the hardwood frames.

In early 1938 the TVA conducted experiments on their land at Eckernförde-Borby with a rotating base on which a DeTe-II was mounted. The experiments were intended to determine the accuracy of the air-warning radar for range and direction. Von Willisen had outfitted the set with audio generators of 2000 and 1000 Hz, corresponding to ranges of 75 and 150 km respectively. The indicator had a dual-beam CRT with deflection amplifiers, one for surveying the whole range, the other for adjustment of the Messkette. For lobe switching using right and left antennas, signals were displayed with one inverted relative to the other. For exact comparison, the second CRT beam allowed a magnified display.

In July 1938 Hitler, Göring, Raeder, the Chief of the General Staff (elect) Halder and other military notables paid a visit to Fleet Command, and this gave an opportunity to demonstrate GEMA's air-warning radar. Göring, whose girth made entering the air-warning set difficult, to Hitler's amusement, was especially interested in the results of locating aircraft. He-111 and Ju-52 were tracked on the survey scale out to 90 km.

Göring insisted that a representative of the Luftwaffe Technical Bureau had to be brought in. He was extremely angry that the Chief of Luftwaffe Signals, Colonel Martini, had not been invited. Here again was the secrecy shop-keeping of Naval Ordnance, which did not want Martini to learn the status of Navy radar at all. They did not want the Luftwaffe purchasing radar from GEMA. They insisted that GEMA was exclusively their source.

In early 1938 GEMA delivered a Seetakt set for the pocket battleship *Admiral Graf Spee*. Fleet Command had selected Captain Langsdorff to collect practical experience with radar. The set was installed with the help of the assembly shop GEMA had recently established in Kiel. It had a 10 square metre antenna of ten full-wave dipoles each for transmitter and receiver that was mounted on top of the rotatable optical rangefinder in front of the foremast. Inasmuch as naval personnel were not yet sufficiently trained, NVA and GEMA employees were placed alternately on board, serving to train and orient the sailors. Completing the experiences collected in the use of radar aboard ship with its installation on the *Graf Spee* proved useful. The location of targets in combat-simulation exercises was convincing. The high location on the hood of the optical rangefinder ensured effectiveness to 25 km, held to be sufficient for ships' artillery. The Messkette was dimensioned for 40 km, and accurate direction finding with lobe switching was considered but not used, owing to the complication for the operators. Connecting the two halves into a single antenna enhanced the gain and the range. Naval personnel were oriented in a relatively short time. There were problems with the TS1 triodes, which were not reliable in long, continuous operation, and to change them required an NVA or GEMA technician. This was corrected with the improvement of the transmitter tubes. Erbslöh made a final voyage in early summer 1939 along the south coast of Spain as the last GEMA representative.

On the outbreak of war *Graf Spee* became a commerce raider in the Indian Ocean and the South Atlantic, where radar first proved its usefulness in war. After sinking nine steamers of over 50 000 BRT she came into action against British warships and took refuge in the neutral harbour of Montevideo in order to repair damage. Langsdorff scuttled his proud ship in shallow water on 17 December 1939 as a consequence of a supposed concentration of enemy ships before the mouth of the Rio de la Plata. British radio experts recognized from photos of the wreck the radar antenna and had it examined. At this time the Royal Navy was equipped only with 7.5 m air-warning radar, which had no capability for observing ships, except at very close range.

The demonstration of the air-warning radar at TVA had caused a positive reaction with Göring, which caused a contentious reaction with the Navy. Because of the high state of the GEMA equipment, the Luftwaffe Technical Bureau was ordered to confer with Telefunken and Lorenz in order to determine what, if anything, they had in the way of radar useful for Flak (the antiaircraft arm of the Luftwaffe). Since mid-1937 there had been discussion about methods for determining aiming data for antiaircraft guns, but there had been no tests of fire-direction apparatus, although there had been a few laboratory experiments.

In order to evaluate the new field of radar, a special command was organized at the Flak Artillery School III at the Lynow firing range in autumn 1938. To make a detailed study of their capabilities and suitability for fire direction the Luftwaffe ordered radar sets from GEMA, Telefunken and Lorenz. GEMA delivered a DeTe-II for 2.4 m on a rotating base. The Navy insisted that this set not be the latest version and not have lobe switching; the transmitter used grid modulation of the GEMA triodes TS4 and delivered 10 kW; the indicator used a Philips DG16 cathode-ray tube and had a pulse repetition rate of 2000 Hz, hence a maximum range of 75 km. Although the call had been for sets intended for the control of antiaircraft guns, the GEMA equipment brought to Lynow determined direction only by maximizing the return signal. The Messkette that accompanied the set was calibrated for 60 km. Flak designated the set the A1.

Lorenz also delivered a set in 1938, designated the A2, but, in contrast to the GEMA equipment, it was an experimental set designed specifically for antiaircraft guns. Its indicator used a CRT with circular trace for the time base on which the operator positioned a mechanical pointer to the flank of the target echo of the tube trace, and this was arranged to transmit the range to the gun director. GEMA put a similar range transmission into the Messkette. Lorenz's A2 set imitated the GEMA DeTe-I in using 60 cm. With a power of 1 kW it had a range of 8–13 km for aircraft, as accurately determined as in the GEMA equipment.

Telefunken called their contribution 'Darmstadt', designated A3 by the Luftwaffe, and demonstrated it on their own test grounds in Gross-Ziethen. It worked on a wavelength of 53 cm and, with 2 kW, followed aircraft at 10 km. It was originally intended for Flum to track aircraft in detail once they had passed the long-range air-warning radars, but its value for Flak was quickly recognized, and accurate bearing and elevation would be provided later by a rotating dipole that produced a conical scan, in function equivalent to lobe switching.

The careful investigations conducted by the Flak Artillery School established fundamental differences between the A1 radar of GEMA and the equipment offered by Telefunken and Lorenz. Whereas the Lorenz set, and especially the one from Telefunken, had the possibilities for accurate location of the target in three coordinates, GEMA's A1 was able to search for and locate targets at much greater ranges. The A1 was a fully developed model capable of being delivered quickly from production.

Their investigations and tests introduced the Flak School to subjects common to the developers of radar. For the first time Erbslöh and von Willisen had the opportunity to discuss the concepts and requirements of the Luftwaffe with regard to defence against fast fighter planes. In January 1939 at Lynow, Erbslöh commented by chance to men from the Luftwaffe Technical Bureau, while converting the A1 base to motor drive, that the equipment should be enhanced so as to function as a set for panoramic search. The set being examined should be outfitted with a polar-coordinate CRT indicator, which GEMA had tried earlier for a circular-range trace but had discarded in favour of the Messkette. The time of rotation would be accomplished with magnetic coils at the neck of the tube that would be coupled mechanically with the rotating base. The time base of the signals would proceed outward in the radial coordinate and the beam intensity modulated by the signal strength. With that Erbslöh established the principle of the panoramic apparatus that GEMA would later erect for air surveillance on the tower at Tremmen, but he was unable to find enthusiasm for the idea, so important for the future. They limited themselves for the moment to an air-warning set of great range for seeking and guiding aircraft.

Since 1937 GEMA had worked on another idea of Erbslöh's intended to solve the problem of identifying the ship or airplane from which the echo returned. These must carry a device that allowed the interrogating radar to identify them. Such devices were called 'Kennungsgerät' (recognition apparatus), or IFF in English (identification friend or foe). The IFF had a receiver that accepted the pulse from the interrogating radar and triggered its transmitter to key a coded response. At the radar the signal returned from the IFF would be received and identified. GEMA had tested this system many times in their air-warning equipment in the region between Eckernförde-Borby and Pelzerhaken, and it found strong approval with the Navy. For such a system to work for both ships and aircraft naturally required the Navy to discuss the matter with the Luftwaffe and to agree on a common system. A meeting was held at the end of 1938 when the Navy learned for the first time that the Luftwaffe was going its own way with Telefunken and Lorenz. No understanding was reached governing a common IFF.

GEMA did not allow this to divert them from their own IFF work and continued it for application to DeTe-II. This required, of course, the use of tubes suitable for aircraft, and it was difficult to find a good tube for the IFF transmitter. Provisionally the RL12P10 was used, which was only usable for a minimum wavelength of 3 m. Philips wanted a contract to develop a tube for GEMA that would deliver 1 kW on 1.5 m.

There was, however, not enough time in 1939 to test the DeTe-II as a panoramic device, although there were concepts for two forms. The first would have swept a sector as the base swung back and forth with CRT display in the normal manner. The second would have had the base rotate continually through 360 degrees and displayed the echo picture on a television tube. Both designs were later completed. When combined with IFF they constituted the feature of modern surveillance radars that ensure safety for both air and sea transport.

CHAPTER 13

THE TIME JUST BEFORE THE WAR

The number of GEMA employees had grown to nearly 800 by the end of 1938, but there were still not enough to meet the rising requirements for sonar and radar sets in addition to other obligations. To the great displeasure of the Navy GEMA now counted the Luftwaffe among its customers, which, given the circumstances, could only bring their delivery schedules into disorder.

In September 1938 Britain and France agreed to ceding the Sudetenland of Czechoslovakia to Germany, following a policy of appeasement toward Hitler. Tension arose immediately between the residuum of Czechoslovakia and Germany, which Hitler used in March 1939 as a reason for German troops to invade. The tension accelerated the procurement of arms and a heightened demand for new weapon technologies. Immediately the Navy, followed quickly by the Luftwaffe, pressed to obtain their orders, and the Luftwaffe gave notice of the need for 250 air-warning radars.

GEMA was able to settle the radar frequency bands that the Navy and Luftwaffe would have to share after a great deal of effort. The air-warning band was assigned a wavelength of 2.4 m, corresponding to frequencies between 120 and 150 MHz and designated with the letter 'f', whereas the Seetakt band was assigned a wavelength of 80 cm, corresponding to frequencies between 335 and 430 MHz and designated by the letter 'g'. These letters would be used to identify equipment as DeTe-f and DeTe-g until the final troop nomenclature was formulated later. GEMA used both bands in stationary equipment with rotating bases. The DeTe-f with two arrays of six dipoles each for transmitter and receiver was at the disposal of the Luftwaffe for air warning. The Navy received the DeTe-g, also available with a rotating base, for use as coast watching and as individual units for service aboard ship. Its antenna was the same but given a mounting suitable for ships; it had ten full-wave dipoles each for transmitter and receiver and had provision for division into two halves for lobe switching, if that was called for later.

In January 1939 there were seven DeTe-g sets ready for service but without lobe switching, which the Navy had still declined to use. These sets were intended for installation in the new battleships *Gneisenau* and *Scharnhorst* as well as the new heavy cruiser *Admiral Hipper*. Preparations at the shipyard had reached the stage that the radars could be installed on short notice. All antennas were so constructed that lobe switching could be added whenever this modification was called for. By September 1939 there was a total of 104 DeTe-g sets ordered for installation on battleships, cruisers, destroyers and torpedo boats as well as for gun- and motor-boats. The installation possibilities and preparations required extensive discussions among the technically competent staff of the Navy and its shipbuilders.

In January 1939 there were six DeTe-f sets ready for the Luftwaffe. In the GEMA shops work was pushed on the rotating bases, together with Zeiss-Jena. Under the press of circumstances the Luftwaffe took away beforehand two rotating bases with antennas and electronics. Because of the international situation they set these up for observing the airspace in the direction of Czechoslovakia with one set on the Geisinger Berg near Altenberg-Erzgebirge and the other on the Grossen Schneeberg (mountain) near Glatz-Sudetenland. Both sets were assembled by GEMA personnel. The two sites were selected because of their 800 m height in order to observe the airspace to the greatest possible range. The operation of these two such secret devices was hindered by the camouflage measures required and which the GEMA mechanics experienced for the first time, but a more substantial problem was found in the many puzzling echoes that appeared on their indicators. After a few days they realized these resulted from the mountainous terrain in which they found themselves. The high altitude of the sets proved to be very unfavourable because of the large number of ground returns, and the echoes from aircraft could be identified and evaluated only with effort. Even von Willisen with his great experience with radar, and who was called on for help, could not contrive any improvement. He suggested observing the Czech airspace from the floodplain east of Vienna.

It was to GEMA's advantage in solidifying their relationship with the Luftwaffe that they made a set available within two days for transport to Vienna. The bulkiness did not permit the transport of an assembled set on the road at this time. The base had to be essentially taken apart and reassembled at the location. Within a few days the radar was in operation, and the result was incomparably better. The airspace in the direction of Brünn could be observed without the clutter. The construction of a fourth set in the Danube plain near Passau was not completed, as Hitler occupied Czechoslovakia in March 1939. (Metre-wave ground radars are best sited in a shallow bowl of land. This restricts ground returns to nearby objects that are easily recognized and allows aircraft at much greater distances to stand out.) In addition to the tests at Lynow this action procured for GEMA a breakthrough for their equipment with the Luftwaffe.

GEMA did not acknowledge that they had or would have a serious competitor, because the Telefunken and Lorenz sets could not attain the distances required for air warning, although they were to prove suitable for Flak. Telefunken's designers profited from the pulse techniques they had employed for television.

Erbslöh and von Willisen concerned themselves with the extensive hiring of good scientists, engineers and technicians to enhance the competence of the company. The more the international political situation sharpened in 1939 the greater became the pressure from the Navy, and now from the Luftwaffe, to begin expanded production on the basis of current designs.

Naval Ordnance continued to be unhappy that GEMA was making deliveries to the Luftwaffe. Their contract gave them the right to participate in, or at least be informed about, the discussions being held with their sister arm, and they used this right consistently to preserve their own interests and give their own needs higher priority. Owing to the continuing pressure from this prime contractor, and fearing bottlenecks in production, GEMA began allocating some responsibilities to subcontractors, despite the difficulties imposed by secrecy. They secured the controlling interest in a foundry, the Rastatter Metallgiesserei, which assured their needs in cast aluminium chassis, used in almost every electronic unit they manufactured. These parts determined the mechanical stability and high resistance to corrosion of their products. Preparations were made with the firm of Metzenauer & Jung in Wuppertal and in Heilendorf-Sudetengau for production of complete power supplies for all DeTe transmitters. Another subcontract was negotiated with the firm of Voigt & Haeffner for complete sound generators and power supplies for sonar. This equipment contained no essential features that had to be kept secret and were allowed to be fabricated outside their plant.

Still, these measures did not suffice to ensure deliveries to the Navy in the quantities on which they had planned, given the Luftwaffe's additional appetite, and the Navy threatened GEMA with awarding rights to AEG for manufacture. This demand must have astonished GEMA, because the Navy had forced them to cancel in 1935 their contract with Lorenz that was intended just for the purpose of utilizing that company's plant. AEG was a traditional supplier of technical equipment for the Navy and therefore stood in high favour. It was not out of the question that the Navy did this out of anger at GEMA's new business connections with the Luftwaffe rather than out of benevolence towards AEG in the hope of an increase in the production of GEMA equipment.

GEMA had encountered AEG on several occasions through its work with the Navy, especially in their factory on Drontheimer Strasse in Berlin, their operations centre for naval, aircraft and army construction. For some time there had often been the need for co-operation during the planning for and the installation of radar in ships, and this had gone amicably. In the naval contracts AEG was responsible in their plants as well as in the yards for special technologies that were necessary for the installation and operation of GEMA equipment aboard ship. They had the necessary assembly capacity to install electrical current and control equipment for radar and sonar. GEMA readily accepted AEG co-operation, as they were not and did not want to become competent in such broad areas, but they were not in agreement that AEG should obtain knowledge with which they could develop their own sound- and radio-location apparatus or further develop GEMA's. Long negotiations were carried out with the assistance of Naval Command to determine the relationship between GEMA and AEG and to protect the patent rights of both firms. Nevertheless, as a consequence GEMA fell into the undesired legal position that they had to let AEG learn, even though to a limited degree, the technical details of their equipment. During this work GEMA did not want to hinder AEG from improving and further developing GEMA equipment. GEMA had to concede AEG's own claims, the use of which they could not obstruct after this construction period. The collaboration between the two companies was regulated in a contract that required Naval Command to clear up the difficulties and disputes that were to come.

AEG had Technisch Physikalische Werkstätten (technical physics work-shops), abbreviated TPW, in which they developed cathode-ray oscilloscopes and their applications, among other things. GEMA had given the firm of Loewe a contract to produce the Type GEB Indicator, stipulating delivery with their own CRT or with the Philips DG16. AEG-TPW had developed a dual-beam CRT with 100 mm diameter screen. In a discussion in October 1939 GEMA determined that AEG should quickly develop for them an indicator unit with one or two dual-beam CRTs and demonstrate it by January 1940. Thus from work undertaken together in part was born the celebrated dual-beam indicator for GEMA radars. This compact device, in which two or four electron-beam systems could be encompassed, allowed GEMA engineers new possibilities for improving the accuracy and versatility of their sets while simplifying operation. The co-operation with AEG-TPW paid off. It paid off for AEG as well, for they became the exclusive manufacturer of indicators for GEMA radars.

A close co-operation with AEG on Drontheimer Strasse was thus established. Together with the solution of all the problems having to do with the installation and control of ships' radars, AEG took over from GEMA the development of stationary bases for ground radars, to GEMA's great relief, a task that one of their sections, Elektroantriebe (electric drives), could solve much more easily than GEMA.

By mid-1939 GEMA had delivered the first series of nineteen sonars to the Navy. Some of these sets were sent to the ultrasonic sound laboratories in Wil-helmshaven and Kiel for testing and inspection. Thirteen sets were tested aboard various ships and submarines. With these tests came, on the one hand, experience in the tactics of deployment and, on the other hand, they provided possibilities for training the operating personnel with the equipment. GEMA personnel often attended these tests, which were at times conducted under realistic conditions, and obtained information directly about what could be done to improve design and construction. This information reached the relevant technical offices in Berlin by the fastest routes, so that the delivery of the last sets of the first series marked the introduction of an improved model into production. After August 1939 GEMA produced a sonar of outstanding design and construction. It had the unique, pur-poseful aspect that stamped the appearance of GEMA's radars. The disadvantage was that the set was large and heavy, which corresponded to demands of the technical outfitting of ships.

No difference was made between the sonars intended for battleships, for which space played a small part, and for submarines, for which space was a rare commodity. The problem of mounting sonar was particularly difficult in U-boats U-77 to -82, U-88 to -102 and U-112 to -118.

GEMA Type 200 sonars soon replaced earlier sets having Generator HS15 and Console HM15.

The Type 200 consisted of an A-unit with air-cooled generator SA with driver stage SAV, output stage SAE; B-unit with rectifier SB with power supply SBN; G/H-unit with location indicator SG with upper console (half-silvered-mirror indicator) SGK, lower G/H unit (twin amplifiers) SGZ, lower G/H unit (power supply) SGN.

All three modules of the G/H unit were, as were those of the generator unit, formed from cast aluminium. Also as with the generator unit, connections followed self-attaching techniques. Modules could be removed and inserted without the need for removing cables. The front panels of the cabinets for the generator and G/H unit were made of cast aluminium. The covers were equipped with quick-change locks that could be bolted and that actuated voltage interlocks. With the Type 200 sonar GEMA introduced a construction that found high regard. The excellent operation and high reliability was proved in many sets.

It is not possible at this time to learn exactly how the distinctive GEMA cast aluminium construction came about. With the Type 200 sonar, Shop Superintendent Mayrczak employed a form of construction that was so comprehensive that it was then used for all equipment. The techniques for connections and modules were improved again and used in later sets. The principle of modular construction was incorporated in all sets in order to ease fabrication and simplify maintenance, although the first sonars did not have it. A coherent portion of the total circuit was incorporated into a module, and these were joined together to form the whole set. In this manner interchangeable components were available that could be plugged into equipment as needed. A repair often required nothing more than the replacement of a module. This allowed personnel with little technical training to maintain equipment and simplified the stocks of spare parts.

Since May 1939 GEMA had owned 10 000 square metres of vacant land on the opposite side of Wendenschloss Strasse. Originally it was planned to use this space for checking the assembly and disassembly of sets and their rotating bases. The rather open view of this land caused GEMA's customers to allow such work only under conditions of the strictest secrecy, so 15 000 square metres was obtained in the vicinity of Jüterbog, which allowed work to be done under less constrained conditions. Experimental equipment could be constructed and tested under perfect conditions. Barracks were built for quarters and workshops.

CHAPTER 14

GEMA IS BOUND TO ARMAMENTS

The crisis during the summer before the outbreak of war in 1939 increased the pressure by the Navy for production by GEMA, but now the Luftwaffe wanted an accelerated production of radars too. Along with this came major requirements for the training and orientation of personnel on the equipment being delivered to both Wehrmacht branches. While GEMA set up a training centre on Wendenschloss Strasse across from the plant, the Navy built a radar school with GEMA support in the so-called 'BG-tower', next to a testing ground for radar equipment at the Navy Docks at Kiel. This allowed their operating and maintenance personnel to be trained further after their basic introduction in Berlin. This tower had served the Navy of the Kaiser and later of the Republic for the adjustment of optical equipment. Such optical devices for heavy naval artillery, made primarily by Zeiss, were by then adjusted and tested in a modern-equipped building on the mole of the Naval Arsenal. After the beginning of 1939 the top floors that had been made free in the tower served Communication Department VIII of Kiel Naval Arsenal for testing. The radars were installed on the seventh of the total of nine floors, surrounded by glass walls. The antennas were mounted outside on the tower with their radiation patterns directed toward the Kiel Ford and the open sea. A lighthouse 15 km away at the entrance to the ford served as a calibration mark.

The sailors on the tower at Kiel, quite different than at the GEMA training in Berlin, had the opportunity to use radar under realistic conditions in tracking ships and to compare their radar measurements with optical results, when visibility allowed. Training airmen of the Luftwaffe was not so easy. A rapid basic course was indeed guaranteed by GEMA, but a full operation of a complete ground radar could be done only on the restricted space of the grounds adjoining the plant. This led to the Luftwaffe creating their own training facilities faster than the Navy. They set up their own testing and training stations as well as that of Army Ordnance at Lynow-Mark, just as had been done in November 1938 at the Flak

School III. It was, however, not possible within the short time available to form a sufficient number of operating personnel and officers, whose activities required knowledge of the use, installation and function of radar.

Already at the beginning of 1939, as re-armament was accelerated because of the Sudeten crisis, GEMA had increased the production of DeTe-I and DeTe-II. All that retarded the general use of radar in both Navy and Luftwaffe was the training programme that had begun too late. The Navy and Luftwaffe could have had more than fifty radars in operation at the start of war, if sufficient operational personnel had been ready. But at that time neither the relevant organizations nor the authorities for training had been formed.

GEMA could for the moment extend production fairly easily. They did have to be sure they filled the contracts for sonar, which had enough urgency to affect the radar work being done for both Luftwaffe and Navy. GEMA radars were not employed in the war with Poland. By that time more than fifty sets had passed inspection, but they could not be employed because of the lack of qualified operators.

After the invasion of Poland on 1 September 1939 Great Britain and France declared war on Germany two days later as the consequence of their responsibilities to Poland. Now Berlin, and especially the area in the west of the Reich, was threatened by possible attack by enemy bombers. Already in 1938 the border from Switzerland to Aachen had been reinforced with fortresses. Beyond this so-called 'Westwall' a 100 km deep air-warning zone was planned, based on the capabilities of DeTe-II, and a number of sets were ordered. The Luftwaffe was responsible for Warning Zone West from the Swiss border to Ostfriesland. The Navy had planned their own stationary DeTe-I and DeTe-II equipment for the islands of Borkum, Helgoland and Sylt. Installation of air-warning sets for the Black Forest, the Pfälzer Wood, the Hunsrück and the Eifel was to begin in 1940. The naval installations on the islands to protect the Helgoländer Bucht was to begin operating by the end of 1940. The outbreak of war altered all these plans.

In September 1939 Erbslöh and von Willisen were informed about how their corporation was to fit into war production. GEMA was declared the prime Wehrmacht contractor for radar. As they alone had models ripe for production, they received the proclamation with great seriousness, for with this measure GEMA became an important organization in matters dealing with radar. Their work had priority and must, according to an order of the Führer, be especially favoured in the distribution of material and labour. This proved to be a bitter experience for Erbslöh. While his partner, von Willisen, was required by his duties to organize the radar infrastructure of the Navy, Erbslöh had to establish himself as business manager of the company. He had to place himself behind the hard war conditions imposed by his contractors, which overruled his humane thinking.

The pressing requirement for rotatable bases was covered with a new series that replaced the obsolete hand-made wooden construction. They did not yet have the newly devised antenna mounting frame that could be hoisted into position hydraulically. Also missing was the motor drive for lateral motion, so direction could be set only with a handwheel. The electronic units had to be set up after the assembly of the rotary base. The final plan had a rotatable base that was mounted permanently on the carriage and that, except for the transmitter, housed the electronics. The profile of this base was so dimensioned that with the antenna support lowered and folded the whole could be transported on a trailer pulled by a tractor on the roads. In marching order this new base, designed in co-operation with AEG, had a height of only 3.6 m. In operating condition with antennas raised, it had a height of 8.5 m. For lateral motion there was a geared direct-current motor, which was driven by a Leonard rotary converter. This device had a motor generator for converting the 380 V three-phase current into direct current whose sign and voltage could be controlled, thereby determining the direction and speed of the dc motor. Inside the base were two handwheels as well as two position indicators. Later the lateral-motion drive was altered to have a special selsyn for transmitting bearing data. The whole drive and its control consisted of AEG components and construction. AEG used this drive aboard ships as well, which permitted assemblies to be made that were common to the air-warning and Seetakt sets. AEG also became the supplier of the complete turning gear. The cabin was fixed to the pivot and rotated by hand or motor. For bringing the necessary connections to the base AEG furnished a slip-ring assembly that was built into the rotating base of the carriage. The cabin, which had previously been made of a partially round wooden-plank construction with a single door, was replaced with an all-metal rectangular construction having two doors.

In order to place the antenna as high as possible above the ground, it was attached to a frame that was fixed to the base and that could be raised hydraulically, either by hand or by motor. The firm of Christoph & Unmack developed for GEMA a special antenna wagon on which the unwieldy 6.5 m by 2.5 m antennas could be loaded for transport. The antenna arrays were formed from frames of metal tubing braced with tight wire and on which the supports for the dipoles were mounted. The transmitter and its cooler were mounted adjacent to the antenna that they served. The receiver antenna, which looked like that of the transmitter, had a Lecher wire with an adjustable short-circuit bar that moved longitudinally in the air-warning set, rotationally in the Seetakt. This allowed optimum matching of the antenna to the receiver cable.

AEG developed and delivered a compact unit for housing and transport, at first with a Leonard rotary converter, but later with a stabilizing three-phase rotary converter complete with switchboard. The same machines were used for Seetakt sets aboard ship. For Seetakt air-warning sets they had housings with hinged walls. They were so constructed that the generation of voltage for the rotatable base required only the connection of a cable from the three-phase mains or other suitable current source. The connection to the slip rings of the rotatable base also made use of a special cable.

With the help of AEG, GEMA's engineers were easily able to assemble a radar for mobile air-warning use on land, light enough to guarantee rapid changes of position. This supported ideas prevalent among the military at the start of the war regarding the expected deployment of radar for air-space surveillance and combating enemy aircraft. Rapid change of position was possible with a motor train consisting of a tractor pulling the trailer and the radar and a truck pulling the mobile generator and carrying mechanical units of various kinds. The antenna wagon was also pulled by this truck. The Navy liked this lightweight set too and ordered units for 2.4 m and 80 cm. In order to make transport by rail possible, GEMA constructed a base that could be separated from the pivot with a crane and loaded on a separate rail car, thereby conforming to railway profile requirements.

Other than the high-priority Seetakt sets planned for warships, GEMA provided the Navy and the Luftwaffe with some urgently requested radars, here and there with improvised and incomplete bases. By the end of 1940 they had assembled with the help of subcontractors, either outdoors or in the new hall, 17 bases. Nine bases with DeTe-II sets were supplied to the Luftwaffe for air warning on the west border and for the defence of Berlin. The Navy received a DeTe-I and DeTe-II pair for the islands of Helgoland, Sylt and Borkum. The Navy took over the remaining two bases with DeTe-II sets for interchange and for instruction. All of these were furnished, according to the wishes of the contractor, only with the elements needed for ranging and for direction determined by maximizing the signal. Although GEMA had a proven design for the technique of lobe switching, they found no interest in the people responsible for ship and antiaircraft artillery. Flak held out for the equipment intended for their special needs then under development at Lorenz and Telefunken.

In autumn 1939 GEMA put into operation on an 18 m tower in the testing ground on Wendenschloss Strasse a radar using the old 61 cm Seetakt wavelength that they proposed as a gun-laying set. For direction the signals of two continuously switched receiver antennas could be compared on a CRT screen either next to one another or over one another. Although sensitivity was sacrificed by using only one half of the antenna array at a time, aircraft over Berlin were located with an accuracy of 0.2°, if the echo could be extracted from the noise. A similar set mounted on a carriage failed to arouse interest for Flak. Nevertheless, they developed this set as 'Seeart-Anlage' for direction of artillery fire from ships or coastal artillery positions. For it were devised the antenna lobe switch ZSU2 and indicator ZSB2 with cathode-ray tube DG9. Flak had quickly recognized that DeTe-II was, because of the distance at which it could pick up targets, outstanding for air-warning service and that it would fit nicely with the gun-laying radars in development elsewhere. This definitely propelled GEMA in the direction of equipment having very long range capability.

Various combinations of DeTe-I were available in March 1940 for the Navy on ships or on carriages. Similarly DeTe-II was available for the Luftwaffe, of which Flak was a branch, for air warning. **The various configurations were transmitter T with ultra-oscillator TU or GSU and control unit TS or GSS, receiver N with input NE or GEE and phase-shifter for T (GEP), observing unit O with Messkette GOK/GOKL and indicator GOB with a 16 cm tube**

DG-16, control and power supply GB/R wherein the audio oscillator was occasionally incorporated, transmitter and receiver antennas V and W with matching unit WA, and equipment providing electric current for the set and for lateral motion.

Of the 17 carriages, 11 could be transported complete; six required auxiliary transport and had to be assembled on location.

Varied climatic conditions led to the expectation of reduced reliability, as had been learned from experience aboard ship. To ensure the secure coverage of the western border with seven air-warning sets, GEMA sent, at the behest of Flak, personnel from the shops at Kiel and Wilhelmshaven as well as technicians from Berlin to these advanced stations. It proved, as had been found before, to be very useful for development people to collect experience in the field. Tubes that had to be worked at high voltages were weak elements. The lifetime of the hard-driven transmitter triodes was less than 200 hours, which demanded new designs that would incorporate cathodes capable of extended service. To strengthen their own tube production, which now had more than 60 workers, GEMA commissioned AEG to manufacture the newly developed high-voltage rectifier VH3 and the further developed transmitter triode TS4, which in a later version was called TS41. These transmitter tubes proved themselves over the next few years capable of generating 35 kW in grid modulation and up to 200 kW in anode modulation. The TS1, developed for the 60/80 cm band, had to be replaced by the TS6, which was then in development, in order to enhance lifetime and reliability.

At a time when Erbslöh and von Willisen, with a group of close associates, were considering the problems of the civilian use of radar for navigation of commercial shipping and aircraft, there were more than 20 GEMA radars on land and at sea for military use, and 20 more were being prepared for installation in the Navy. Series production of sonar was in full swing and by the end of 1939 the Navy had accepted 37. On 31 December 1939 GEMA employed a total of 1643 workers, of which half were for production. In order to fulfil the many orders for sonar and radar, a marked increase in production was required for 1940. In order to ensure the proper maintenance and operation of the sets in the hands of the customers it was necessary to develop and provide a wide range of test equipment that was not obtainable from the electronics market. To this purpose a group of 200 people was formed. For the management of subcontracts awarded to AEG, Metzenauer & Jung, Voigt & Haeffner, Danner, Schaub and Blaupunkt, an administrative section, Büro für Auswärtsfertigung (BAF, office for external production) was organized, which had the full pertinent responsibility for procurement of such items as rotatable bases, generators, mechanical components, modules and equipment which the requirements of secrecy allowed. Deliveries through BAF, which grew substantially as the war continued, functioned essentially the same as deliveries from GEMA plants. GEMA accepted the responsibility for all products, regardless of their origin.

The collaboration of the detached GEMA technicians sent to help establish the seven air-warning sets on the western border made possible nearly uninterrupted observation of French and British aircraft in that airspace. With help from the NVA the Navy built stationary Seetakt and Flum sets on Borkum, Helgoland and Sylt for controlling sea and air traffic in the Helgoländer Bucht. The Luftwaffe had given to the Köthen Regiment (Nachrichten-Versuchsregiment, a regiment devoted to experimental radio techniques) one of the DeTe-II sets that had been used during the invasion of Czechoslovakia. With this set Lieutenant Hermann Diehl formed a research group on Wangerooge, which had a spectacular success on 18 December 1939 when used against an attack on Wilhelmshaven by British Wellington bombers. With his radar he was able to locate the RAF bombers and guide German fighters over the North Sea to attack them. With this loss of 12 machines Bomber Command first encountered GEMA radar—unsuspected and unrecognized. Diehl perfected in the following months the first German radar-controlled fighter direction. Through his initiative grew the techniques that became significant when the attacks came in the night, the result of unacceptable losses during daylight raids. However, the absence of a method for determining altitude with radar hindered the effectiveness for this new night fighting.

For a rough experimental determination of altitude Diehl raised and lowered an antenna array with horizontal dipoles on masts as high as 20 m, but the success did not justify the expense. GEMA took part in these experiments only indirectly but followed the work done with their sets by the Köthen Regiment with interest. Given that the radars being developed specifically for Flak determined altitude, GEMA concentrated instead on extending the range of their sets.

On their grounds at Jüterborg in early 1940 GEMA began the first experiments in attaching a number of DeTe-II arrays on top of one another to a mast. The first experiments showed a doubling or even a tripling of the maximum range over a normal air-warning set. Here began the realization of rash dreams for air-warning radar. With such antenna configurations and increased power it became possible to pick up RAF aircraft as they rose from their bases in England. For this approach GEMA found immediate support with the Luftwaffe, which was quite interested in air-warning radar of greater range.

As the interest of the Luftwaffe for GEMA products grew, their relationship with the Navy deteriorated. The Navy had a direct view of this business through their Contract Office in the GEMA plant, which furnished the administration for Luftwaffe deliveries too. At the beginning of the war Naval Ordnance required formal meetings between them and the Contract Office about production. At these meetings production reports for sonar and radar were examined. Naval Ordnance had the plant administration on a leash, so to speak, and attempted to influence production for their own ends. After the start of war this produced a negative effect on research and development. Through the great demands for production and through the necessary transfer of valuable technical people from research into the installation and operation of sets in the field, there remained hardly any capacity for the recognized need of further development of the product.

At the beginning of 1940 Erbslöh, and especially von Willisen, planned to separate the laboratories from the mother company and make them independent in order to enhance their efforts and to given them greater freedom. They were ready to found a new corporation in order to escape the situation into which they had slipped as the prime Wehrmacht contractor for radar. Their purpose was to restore the research and development climate that had characterized GEMA in the beginning, and to that end they were willing to give up some of the profits that were coming to them unexpectedly as armament manufacturers of radio- and sound-location equipment. Both saw the danger that the NVA, which was steadily expanding, would determine development, leaving GEMA only responsible for production, and this did not appeal to the two engineers at all.

In forming a new, independent laboratory, difficulties arose both from the nature of the project and from its financing that presented Erbslöh and von Willisen with serious obstacles preventing them from reaching their goal. It could only be done if a significant number of qualified engineers and technicians were hired, but this was no longer possible. The contractors had absolutely no understanding at that time for a diversion of personnel from production to research and only increased their pressure on the company. Erbslöh and von Willisen's dream was simply not to be realized. They would have liked to enter neighbouring fields of sound and radio location, covering the needs of new customers. They did not resign themselves to renouncing research, however. Even if their contractors were not happy, they instituted new developments that secured for them a competitive position with respect to other firms.

The reader may, by this time, be puzzled that the German Army (das Heer) has not been mentioned. In the Wehrmacht the Luftwaffe provided antiaircraft artillery for all except naval use and the Navy provided coastal artillery, two branches of the services that belong to the Army in both British and American practice. Thus the German Army had no radar.

CHAPTER 15

WAR DOES NOT STOP RESEARCH

The war year 1940 found GEMA's research and development in top form, expressed in over 100 patent applications, certainly covering the field of radio location well.

Despite the infamous order against research—even with a Führer Directive given by Göring behind it—Erbslöh and von Willisen had GEMA start new projects. (This refers to an order given by Göring for Hitler on 3 February 1940 that military research not completed for 1940 or shortly thereafter should be discontinued.) Many people qualified for such tasks had to be taken for necessary activities in manufacture, testing and supporting equipment in the field. For workers who were or who might become active in the field, the question arose of what their status would be if, when serving with the Navy or Luftwaffe in combat, they became prisoners of war. They were to wear arm-bands on which was inscribed 'Deutsche Wehrmacht', if they came into a dangerous situation, and were given some calming assurances that this would protect them from the unattractive fate of being held as irregulars. Fortunately, a need for these arm-bands never arose.

On 7 February 1940 the Luftwaffe ordered, under the highest degree of secrecy and with the utmost priority, five and later 12 DeTe-f sets with demountable bases that could be shipped by aircraft. This order came as a part of the preparation for the operation carrying the codename 'Weserübung'. In order to head off British plans to land in Scandinavia, Hitler ordered the occupation of Norway and Denmark. Scandinavia was strategically important for German as well as British war plans, as its west coast allowed the control of much of the North Sea and the important shipping routes of the North Atlantic.

Through this order GEMA DeTe-f set serial number 18 became radar-historically famous as the first of the air-transportable sets to be flown from Hamburg-Fuhlsbüttel to Stavanger in the first days of May 1940. The set was assembled on the coast of Stavanger under the direction of shop superintendent Fritz Henke, who with von Willisen during the war often attended to the highest needs of the firm. It was the first German radar to see effective service there. From its position, Allied ships and aircraft could be observed and engaged. Southern Norway was protected by additional DeTe-f sets brought in later by air. Other sets were installed at Mandal near Kristiansand and at Bergen.

The demountable bases, which were transported as prepared packages and assembled at their destination, consisted of a pivot and bearing for hand operation. This stood on a cross support with a beam fulcrum. The wooden cabin had a welded angle-iron frame attached to the base. The dismounted wooden frame for the antenna was fixed to the cabin in which were found electronic units R(GR102/1), T(GT100/1), N(GE241) and O(GO109/1).

To simplify operation Messkette (GOK) was not incorporated, but Messkette appendage (GOKL60) allowed an increase in range from 60 to 120 km. This sufficed for range on the screen in two ranks, from 0 to 60 km and from 60 to 120 km. The high reliability that Henke had assured by providing adequate replacement parts impressed the Luftwaffe operating personnel, well-trained soldiers who came from the Luftwaffe testing laboratory at Rechlin. After the occupation of Norway, Scandinavia became a strategic base for the Wehrmacht, which required a chain of radars be built along the entire coast both for sea search (DeTe-I) and air warning (DeTe-II).

While there was the Phoney War (in German 'Sitzkrieg') on the west border of the Reich during the winter of 1939–40—the feared attack by the Allies not taking place—the Navy scored some notable successes over the Royal Navy. Battle extended over the North Sea and included the entire North and South Atlantic, where merchant shipping was hunted down. In these actions large and mid-class surface ships took part, but especially feared and effective were U-boats, which disposed of large numbers of enemy naval and merchant ships. In order to make rapid use of their technical superiority with sonar and particularly radar, Fleet Command demanded the speedy delivery of Seetakt sets. GEMA also developed for U-boats, until then only the recipients of their sonar, a special radar, DeTe-U. The severely confined space aboard a submarine made a reduction in the size of the electronics necessary. An azimuthal sweep of 25–30° in front of the conning tower was not made with a movable antenna but electrically with phase-shifters in the transmission lines to an array of dipoles that were rigidly fixed to the front walls of the conning tower. At the behest of the re-organized NVA, now the NVK (Nachrichtenmittel-Versuchs-Kommando), GEMA delivered two prototype DeTe-U sets for installation and trial in April 1940. As a consequence of the meager transmitter power and low antenna location, the maximum ranges were only 5 km for ships and 10 km for aircraft.

By mid-1940 38 sonars were delivered to U-boats of pre-war construction and installed in some; 91 units were delivered to Navy shipyards for installation in new vessels or in those already in service. Since the start of fabrication of the new series of sonar for surface and underwater vessels in August 1939, there had hardly been any change in design concepts, and after intensive employment no

94

basic deficiencies had appeared. Installation offered no problems to the trained shipyard personnel, but finishing quickly the 370 sonars yet to be delivered required an extensive expansion of production.

While Schultes put pressure on his high-frequency department to obtain increased performance from DeTe transmitters and receivers, Brandt and his people worked intensively on the further development of indicator units for both radar and sonar. Even though the higher accuracy of the Messkette had not found complete acceptance with the Navy and Luftwaffe, their ease of operation, and above all their resolution and accuracy, were being continually improved. In practice the operator of the Messkette had to hold the echo pulse on the null mark of the CRT. To this purpose the Messketten of the first design generation, GOK101 for Seetakt and GOK108 for Flum, were provided with fine adjustments, two scales and relay circuits. Four delay elements, A, B, C and D, were connected preceding the fine adjustment. One step of the A element corresponded to a distance of 10 m, of the B to 100 m, of the C to 1000 m, of the D to 10 000 m. The steps of A–C were switched, whereas D used relays.

Geneva mechanisms were used to transform the motion of the hand crank to actuate the switches. Shop Superintendent Mayrczak constructed this switching contrivance so as to allow the operator to continue increasing or decreasing range with crank motion in the same direction without the need to resort to any other switch, which would interrupt his tracking.

The first Messketten were made of coils and capacitors that were aged artificially, then measured and sorted out to accuracies of 0.1% using precision-bridge circuits, but this did not satisfy Brandt. He invented a calibration procedure for the new circuits that made use of trimmer capacitors and coils to set final values. **A typical radar audio-frequency was taken as standard and was caused to deflect a circular-scan CRT. The CRT beam was modulated with a fixed, integer multiple of the deflection frequency, so that the circular trace was observable only as points. The number of points was proportional to the frequency multiplier. The positions of the points corresponded to the relationship in time of the momentary value of the deflection voltage to the trace voltage. A rotation of the trace corresponded to the travel time of the radar signal. The visible points on the circumference represented distance markers. If a Messkette intended for calibration was connected into the path of the circular deflection voltage, then one could, by counting the number of points on the CRT of the calibration unit, establish the distance scale.**

This Messkette calibration unit proved its utility. It served as a precision measuring device for checking radar sets. Already by the end of 1939 GEMA had delivered it in a simplified form for the control of sets that were operating at the workshops and test stations of the Navy and Luftwaffe. The delay-circuit calibration unit consisted of generator MKG 2000/1000, CRT with optical attachment MKO 2000/1000, deflection amplifier MKV 2000/1000 and power supply MKN.

For checking the frequencies and sensitivities of radar receivers GEMA developed and delivered test set GEPR. It was augmented with an antenna-test-transmitter that could be pulse modulated on bands f and g and that served to test the directional antennas of the radar sets. GEMA also provided a power meter GLM for monitoring sets aboard ship, which permitted the output power to be optimized. By the end of September 1939 GEMA provided a measurement kit for their sonars, which checked the most vital functions of individual sets. The kit contained the test unit audio oscillator SPS, phase-shifter SPP, frequency meter SPF and power meter SLM for 15 and 10 kHz.

The phase-shifter allowed the calibration of the indicator by observing the inclination of the vertical Lissajous line on the CRT within an angle of 180°, depending on the momentary values and the polarity of the deflection currents.

The calibrated phase-shifter from this kit proved particularly useful in checking the bow sonar of the heavy cruiser *Prinz Eugen*. Before and after the ship was commissioned on 1 August 1940, GEMA conducted experiments that showed good range even at high speeds, usually limited by the noise from the swiftly moving water. Direction came entirely from the position of the trace on the CRT, which was accurate to 4°. With the kit's phase-shifter, direction was calibrated and checked. Eventually a compensation for direction for this sonar was planned. On 10 October 1940 Naval Command declared the set ready for sea duty and ordered a similar unit for the battleships *Bismarck, Gneisenau, Scharnhorst, Seydlitz* and *Tirpitz* and the heavy cruiser *Admiral Hipper*.

Some test equipment that GEMA had developed for their own use found interest among maintenance groups of the Navy and Luftwaffe. Trained personnel could test and adjust modules of sets with automatic test equipment. The more equipment GEMA delivered the more the customers demanded to have their installation and maintenance people trained and to receive an abundant supply of spare parts and auxiliary equipment.

In February 1940 GEMA began the development of a shop truck to repair radar sets quickly in the field. They used this vehicle, a three-ton Opel, for exchanging modules, which reduced the repair time to a minimum. It was also used to bring units that were damaged or in need of overhaul to repair shops. The truck stored a complete collection of modules with protection against moisture and dust. There were also tools, tubes, instruments and replacement cables. Determining what to include in the outfit was difficult because almost every possible need had to be considered. Often the first radars delivered did not have components interchangeable with later models, having among other things different pulse repetition rates. Thus the Navy's stationary air-warning and sea-search sets had different Messketten. The result was that no single truck could service all kinds of radar. On 7 March 1940 an extraordinarily harmonious meeting took place between relevant offices of the Navy and Luftwaffe. To simplify supply and maintenance a new kind of configuration was to be introduced. Golde and Dünhof represented GEMA and were well pleased with the change.

This discussion freed, without the usual branch duplicities, a path for bringing, at a time of severe production difficulties, the delivery of radars at the newest stage of development. It was an opportunity not to be missed.

In May 1940, after producing about 60 units, GEMA concluded their first series of radars. Whereas shipborne units continued to work at 2000 Hz, ground-based sets used, according to their purpose, either 2000 or 1000 Hz. The continually increasing range, especially for air-warning equipment, brought with it the necessity of reducing the pulse repetition rate to 500 Hz, corresponding to a maximum of 300 km.

The pulse frequency of the modulator and CRT deflection originated in a generator in the R unit. An uncalibrated phase-shifter was located in the receiver for adjusting the zero position of the transmitter pulse. While trying lobe switching, there appeared over and over the need to correct the sweep start on the screens of different CRTs with additional phase shift. Besides that, it was found in practice that neighbouring radars should not have the same modulator frequencies, as this caused them to interfere with one another. On the other hand, there were tactical situations in which there was an advantage in having such pulse rates synchronized.

Brandt's low-frequency section developed, therefore, a frequency unit, the Z-Gerät, with properties for universal application. In addition to the future 500 Hz it could employ avoidance frequencies 494 to 506 Hz in 3 Hz steps. In order to drive several sets from a common source, it had a switch for internal or external generation. In order to adapt the cables necessary for common operation at long distances, both input and output of the Z-Gerät had switches for selecting the proper cable-termination resistance. There was a total of four usage groups in the Z-Gerät. In two groups the phase was coarse with fine adjustment within certain limits. The device was manufactured according to GEMA construction precepts that allowed modules to be interchanged.

Schultes's high-frequency laboratory invented a new concept for the receiver inputs of the Seetakt and Flum radars. These units, bearing the designation NE, replaced the previous single-module receivers with those in which input and IF modules were separate. They were constructed in cast-aluminium housings, assuring mechanical robustness and allowing the individual stages to be thoroughly decoupled and shielded. **The input of decimetre band g of the Seetakt had the well-known Q-enhancing mixer and oscillator stage, both using button tubes 4671. The tuning circuit was a GEMA patent. The coils were circular copper that were brazed onto the ceramic base on which the tube was inserted. Ceramic sectors, adjustable from outside, formed the trimmer capacitors. They were also formed by copper layers brazed to the ceramic with the capacitance altered by changing the amount of ceramic dielectric between plates.**

The input of metre-wave band f of the Flum used button tubes 4672 both for input and mixer, with a 4671 for the oscillator, soldered to special connections on a ceramic base. The oscillator circuit consisted of adjustable capacitors and hairpin-shaped or normal coils. Circuit components that were carefully harmonized with one another contributed to high frequency stability, both for f and g bands, as did their mechanical strength. These

inputs were also used in passive receivers for intercept and testing service; a total of 10 000 were built.

The IF unit became, after changing the tube complement twice, a reliable part of GEMA receivers. The very effectively shielded compartment and the use of 15 MHz for the first IF frequency, which was reduced to 7 MHz in the next, was extremely effective. The use of the sharp cut-off pentode AF100 allowed the bandwidth to be raised to 900 kHz. The instability of the receivers used theretofore had resulted in uncontrolled signal transit times, leading to errors in range in radar sets, which the new receivers avoided.

Brandt persisted in his efforts to improve the most sensitive element of GEMA radars, the Messkette. When the new series of receivers was begun there was a new concept that permitted the construction of a Messkette that was made from fewer components and which was more reliable and simpler to operate. In order to reduce the number of coils, some were made with taps that allowed two range steps with one coil. The reliability of the switches, which were heavily used in operation, was greatly improved by special contacts made of hard-silver and carbon. Data were extracted from the new circuits by way of three adjacent circular scales and a counter. In the standard model the left scale indicated range to 200 km, the middle to 10 km and the right to 1 km. The counter following the kilometre scale had five wheels with two digits after the decimal point.

Field tests of the Messketten had shown that it was difficult to change targets quickly. Changing range using the single crank required too much time and fatigued the operator. A motor-drive could not be considered because such high accelerations damaged the switch contacts.

The new Messkette had two hand wheels, one for coarse and one for fine adjustment. The coarse drive had no influence on the fine, whereas the fine drive altered the coarse by one step with each rotation of the crank. The 1 km step corresponded to a single rotation of the fine drive and was coupled to the 10 km step by a mechanical ratio of 10 to 1; the 10 km step was similarly coupled to the 200 km scale by a ratio of 20 to 1. These two scales allowed themselves to be run back and forth by the coarse drive. The data wheels were so arranged as to allow selsyns for data transmission. With this improvement GEMA offered the Navy and Luftwaffe, both Flum and Flak, the possibility of the electrical transmission of data, either directly or through adapters. This design proved itself and was a feature of subsequent equipment.

Much attention was given to developing highly accurate, long-time-stable coils, capacitors and resistors for the Messketten, which could be made in volume production with reasonable effort. For the calibration of new circuits in production and for control of those installed in finished sets, the low-frequency laboratory developed a test set that was simpler to operate in addition to the calibration equipment described earlier. With this device ('Prüftisch Kette' (Messkette test table), PTK 1) elements of a new or an older circuit needing calibration were checked against standard elements by means of a comparison bridge. In this manner the Messketten could be equalized during construction so that the trimmer capacitors were easier to adjust. Since the Messkette had to operate at 494 and 506 Hz, the elements were set so that measurement error for the lower frequency was −0.1% and for the higher frequency +0.1%. To avoid vexatious cranking for the zero position, an element was embodied that allowed a running control of the calibration over long periods of operation, independent of the range setting.

The first demonstration units provided with dual-beam CRTs gave rise to a new series for general observation and approximate ranging using slow-speed time bases and precision ranging using high-speed traces. **The surveillance unit N contained a tuneable component NA for f or g bands with IF amplifier NZ, and the observation unit NB had two adjacently located dual-beam CRTs. The observation unit made possible the examination of all reflections with the lower CRT. In front of the screen was a scale for the entire range capability. In front of this scale moved a window that indicated the restricted region displayed on the upper CRT, which fulfilled the function of an electronic magnifying glass. The window was moved by rotating the scale-rim of a selsyn phase-shifter, selecting a desired piece of the time base for examination. The range unit O had Messkette OK and the observation unit OB with only one dual-beam CRT, which served as the null detector for the Messkette. At mid-screen was a vertical mark where the echo was positioned by the Messkette.**

Accuracies were achieved with these sets that bordered on the best achievable. In experiments on Köpenick tower and at the company's grounds at Jüterbog that took place in April 1940, the range accuracy attainable was found to be 100 m for targets from 40 to 50 km distant. These results were duplicated in Navy experiments at Pelzerhaken and at the range school at Sassnitz.

The experimental work of Lieutenant Diehl at Wangerooge and the many possibilities opened up by such equipment accelerated the introduction of lobe switching, so much more accurate than determining direction from maximum signal strength. Diehl's introduction of fighter control using a Flum set showed that the accuracy of 1 or 2° so obtained did not suffice. The introduction of the accessory for lobe switching, LPZ, improved the GEMA radar sets chiefly. The attachment LPU provided a motor-driven switching of the receiver antennas, the two antenna halves being connected through a delay line. The receiver outputs were switched in synchronization with the antennas so that the indicator showed the two sets of signals displayed together on the dual-beam CRT, the pulses seen either of the same polarity and displaced slightly from one another, or with one inverted relative to the other. Because the advantages of one or the other display modes were disputed, GEMA developed lobe-switching attachment LPB for both displays. The displaced pulses were observed with horizontal trace, the inverted with vertical. The Luftwaffe found the two displays so good that they ordered both for their Flum sets.

In mid-1940 GEMA introduced a radical change in their production of radar sets. The assertion of the Navy and Luftwaffe that the multiplicity of types made supply extremely difficult was corroborated for the company in the difficulties they encountered when installing radars at the shipyards. GEMA had a stock of 60 module types then in use and from which the Navy wanted first choice. The Luftwaffe contented themselves at first with 22 Flum sets, as the attack on France was offensive and had made little use of radar. Night-bombing attacks on the Reich, given little attention at the start of the war, caused the rapid organization of night-fighting squadrons. Diehl's proven method of directing fighters onto enemy bombers by means of radioed directions based on what he saw on a long-range radar CRT screen quickly took on importance, and the first attempts at radar-directed night fighting grew from his practical experience. Officers, sitting at radar sets, used radio telephones to direct fighters onto incoming bombers. As the number of night attacks grew, Colonel Josef Kammhuber undertook the organization of the night fighters and became commander of the first Night-fighter Division.

Diehl's intercession obliged GEMA in mid-1940 to incorporate the progressive lobe-switching technique into production, although their customers demanded adherence to delivery schedules. They had by then secured themselves an important position in radar relative to their competitors. Lorenz began with the delivery of a small series of gun-laying sets, called 'Kurpfalz', which had two parabolic dishes and which had come from their experiments with the A-2 type radar. The range against aircraft was 25 km with a horizontal- and vertical-directional accuracy of 2–3°, attained from maximum signal. At almost the same time, Telefunken introduced the 'Würzburg-A' for testing, which had a 3 m paraboloid and the same range as the Lorenz set. It also obtained direction from maximum signal strength, attaining horizontal and vertical accuracies of 1.5–2°. Neither set was equipped for accurate direction determination. The Lorenz worked around 60 cm, the Telefunken on 50 cm.

With the start of the new production series, GEMA cleaned up the type designations. Instead of the designation DeTe-I for the 80 cm Seetakt sets and DeTe-II for the 2.4 m Flum sets, the type designations DeTe-g and DeTe-f were adopted, g indicating the frequency band from 335 to 430 MHz and f from 120 to 150 MHz. Internally, however, these designations did not cover the many sub-categories, which required three- and four-digit numbers. The Seetakt sets received numbers between 100 and 199, e.g. DeTe-100, DeTe-101, etc, and the Flum sets between 200 and 299. In time, numbers up to 599 were used for air-transportable sets (LZ), for long-distance sets and U-boat equipment.

At the time of its introduction to the Luftwaffe Signals Troop (Nachricht-entruppe, which was responsible for air-warning) the Flum equipment received the designation 'Freya' by them, suggested by the f wavelength band. When this name became official can no longer be determined. It is interesting to note that Professor R V Jones, leader of the British scientific intelligence service, sought clues into the nature of uncertain information about German radio-location techniques from possible meanings of the cover name 'Freya', the goddess of fertility and sensual love.

By the end of the war GEMA type designations had undergone numerous changes. Only at the end of 1940 was the designation Funkmessgerät abbreviated FMG or FuMG. The cover name, 'DeTe-Gerät', remained in use so long as there were GEMA sets in service. In its diverse forms Freya gained an almost legendary status among the air-warning and antiaircraft troops.

After surmounting the problems associated with initial production, GEMA and its many subcontractors could begin with production of the new series of radar in the fall of 1940. Typical of GEMA's flexibility was that production began a few weeks after the prototype was completed. By the end of 1940 they had manufactured nearly 150 sets of the new series. The number of employees had risen by then to 2500.

Despite the heavy workload in the production departments, which caused laboratory and test people to step in and help, new, progressive radar designs were completed. The IFF work that Erbslöh had instigated in 1937 and that had been patented, which allowed the identification of the target of a radar set to be known to the operator, led to an advanced concept that was ready for production. In addition to the Navy the Luftwaffe showed a lively interest in GEMA's functioning IFF system, although Telefunken already had a system for use with its Würzburg. A demonstration of the GEMA 'ES-Gerät' (echo simulation) with the requisite IFF receiver at the radar set had made a big impression on the Luftwaffe. Such an identification system was vitally necessary for the fighter direction methods then being tested. They ordered some IFF sets with IFF ground receivers for the radars in March 1940 and required delivery of at least 1000 IFFs and 50 ground receivers by the end of the year. In order to avoid an additional set for the g wavelengths, GEMA developed a special IFF that transmitted on the frequency of the Freya IFF receiver. This covered the Navy's concerns with the same ES set as the Luftwaffe, allowing it to recognize its own aircraft.

In competition with Lorenz and Telefunken, GEMA invested significantly in the development of a Seetakt set with vertical as well as horizontal lobe switching. There was still a set with lobe switching for demonstration on the factory tower in Berlin for artillery applications. It was augmented with a vertically directed antenna mounted on the side of the tower for determining elevation. The performance of this GEMA equipment in 1940 was better for antiaircraft gun-laying than those of Lorenz and Telefunken. Because the expense was substantially greater, von Willisen could not find any enthusiasm for it with Flak. When Lorenz and Telefunken greatly improved their equipment, GEMA withdrew for a time.

The alteration of the tower radar for elevation paid off for GEMA, despite the lack of interest by Flak. During the course of this work Schultes tried out a minimum-direction method, which had come to light accidentally in working with a Freya that obtained its lobe switching by means of delay lines. By connecting the two antenna arrays to a half-wave delay line in opposite phase caused a minimum voltage at the mid-point of the cable. This prompted the idea of letting the receiver input connection slide continuously over this delay line driven by a motor, picking up the signal capacitatively. When this signal had a minimum at the middle of the line, the antenna was positioned with the signals of the antenna halves balanced. This forerunner to the later 'Radattel-Peilung', which took its name from the distinct noise of a circulating pickup in a circular delay line, was tried on the tower set and proved quite good. Its advantage over the comparison of two signals on dual-beam CRTs was that it was simpler and that its reliability exceeded synchronized switching significantly. It was easier to operate, as the signal along the delay line and the marked mid-point were clearly observed on the screen of the CRT.

Experiments to extend the range of metre-wave equipment began in autumn 1940 on a 30 m high climbable mast at Jüterborg to which four Freya antennas were attached, which were rotated mechanically through a sector. It could receive a signal from a transmitter of low power on the 10 m tower at Köpenick, a distance of 70 km. In November 1940 the electronics of a Freya were connected to this antenna arrangement and interrogated an IFF in the Köpenick factory. Thereupon the IFF was mounted onto a motor car and driven 170 km to the Brocken in the Harz Mountains to the west-southwest of Berlin. Even at this distance the IFF answered interrogation. These experiments were the beginning of the development of larger radar antennas, giving rise to the long-range early-warning series 'Mammut' and 'Wassermann'.

At the beginning of 1940 the problem arose of making sonar echoes of audible frequencies in order to utilize the Doppler effect for distinguishing stationary from moving targets. This effect, named for the Austrian mathematician and physicist Christian Doppler, increases or decreases the frequency of the signal reflected from an object, depending on whether it is moving toward or away from the receiver. In order to make the small changes in the frequency of the returned signal audible, GEMA developed a heterodyne accessory with which the 10 or 15 kHz frequency of the echo was transposed to 2 kHz. Through this transposition the fractional frequency shift resulting from the Doppler effect was raised by a factor of 5 or 7.5 respectively. The effect of small differences in the target speed was easily heard in either earphones or loudspeaker.

The installation of the large and heavy sonars on U-boats brought difficulties from the beginning. Brandt started the development of a small console in 1940. This unit contained the 'Kino' (motion-picture indicator) SGK, the twin ampli-fiers SGZ and power supply SGN. In addition to reducing the space and weight of these units, it was desirable to control the amplifier gains. Because the echo amplitude decreased rapidly with distance, the amplifier gain had to be increased at long ranges. The low-frequency laboratory found the solution in automatic gain control and achieved it in the new, small twin amplifier, which was also made smaller and lighter using standard Wehrmacht tubes.

CHAPTER 16

HARD BUT SUCCESSFUL YEARS

In the summer of 1940 there arose a situation concerning deliveries that could have become unpleasant for GEMA.

Just as in the Polish campaign, no radar was employed in the defeat of France, as the Luftwaffe commanded the air space so completely that no air-warning system was needed. Because RAF Bomber Command was occupied in altering its tactics to those necessary for night attacks, they made only nuisance raids of limited success over the Reich.

This came at a very convenient time, for GEMA was beginning production of its newly developed radars in fall 1940, and the absence of heavy demand for Freyas allowed them to continue deliveries to the Navy of the new series of shipborne Seetakt sets. This alteration of production fitted well with problems they were having at the time with the bases for ground radar, but the drive for deliveries would soon begin.

The new series had a uniform pulse repetition frequency of 500 Hz, corresponding to a maximum range of 300 km. The multiplicity of types delivered until then was replaced by a standard group. The basic GEMA radar now consisted of power supply R with RH (high voltage) and RN (low voltage), RI (instrument panels), transmitter T with TU (Ultrateil, the self-exciting power oscillator), TS (modulator) and TN (power supply), frequency generator Z with ZP (receiver) and presentation unit N with NE (receiver) consisting of NA (RF unit), NZ (IF unit) and NB (main presentation unit with dual-beam CRTs), fine ranging and presentation unit O with OK (Messkette) and OB (fine range presentation unit with dual beam CRT), matching unit WA, transmission antenna V and receiver antenna W.

Because of their size the transmitter and receiver antennas of the Flum sets were separate assemblies. Each array of six vertically polarized, parallel-fed full-wave dipoles was set before a reflecting screen 6.20 m by 2.50 m. The Seetakt, on the other hand, had a combined antenna (VW). Its transmitter and receiver arrays each had ten vertically polarized, parallel-fed full-wave dipoles arranged before a reflecting screen 3.75 m by 1.90 m.

The design of the 1940 series equipment had so matured that it could be produced and used in large quantities with few problems and few design alterations during the next years, although usage on some sets naturally did force changes. One special version of transmitter T, for example, had to be fitted within the hoods of optical rangefinders, which caused it to deviate from the standard. Radars for U-boats had to have a number of special features in order to be accommodated in the cramped interiors.

On the introduction of GEMA radars among troops the Wehrmacht instituted a new, abbreviated nomenclature that applied to all laboratories and manufacturers. Following the designation FMG or FuMG (Funkmessgerät, radar) came the form of the set: Seetakt (sea search), Seeart (naval artillery gun laying), Flakleit (anti-aircraft gun-laying) or Flum (air warning). Then came two digits specifying the year of introduction and the first letter of the manufacturer, e.g. 'G' for GEMA. A lower-case letter gave the working frequency of the set in coded form, f and g as explained earlier. A final upper-case letter indicated the kind of base and the manner in which the antenna could be traversed.

A = ground, fixed, mechanical traverse
B = ground, mobile, mechanical traverse
D = ground, rotatable base mounted on a bunker
F = ground, fixed, electric scanning by phase control
L = ship, rotatable base on the bridge
M = ship, rotatable base on a mast
O = ship, rotatable base on optical range finder
P = ship, fixed antenna
U = submarine
Z = ground, demountable base

This list also shows the important applications of GEMA radars with the Navy and Luftwaffe.

An example of a DeTe-I set according to the new troop nomenclature: FuMG(Seetakt)41G(gA) denoted a fixed, rotatable base with mechanical traverse; FuMG(Flum)40G(fB) denoted a Freya with mobile rotatable base. This troop nomenclature was extended retroactively to the first GEMA sets. The DeTe-II used in 1938 for observation of the air space over Czechoslovakia received the designation FuMG(Flum)38G(fB). These radar designations were later changed both in the Navy and Luftwaffe to forms that did not provide a description of the device, which was thought to compromise security. From autumn 1943 the Luftwaffe introduced new GEMA radars using sequential numbers 400 to 499. At the same time the Navy introduced FuMO (Funkmessortungsgerät) using sequential numbers 1 to 99 for Seetakt, 101 to 199 for Seeart, 201 to 299 for Flakleit and 301 to 399 for Flum.

The Phoney War ended in the west on 10 May 1940. The German attack began between Luxembourg and Nimwegen with intensive co-operation in the air. On the 15th The Netherlands surrendered, and on the 28th King Leopold III signed the capitulation of the Belgian Army. After the German–French armistice of 22 June, German troops occupied the Atlantic Coast from Dunkirk to the Spanish border, and Operation Weserübung was completed by 10 June, Allied troops having evacuated Narvik and the Norwegian forces having capitulated. Now Germany commanded not only the Scandinavian coast to the Arctic Circle but also the Channel and the Atlantic coast to the Spanish border. Only by comprehensive measures was it possible to defend this region, and GEMA offered a most effective means of guarding this extensive air and sea space with their radar.

Pressure for deliveries followed the intoxication of victory in the west when the requirements imposed by the capitulation of Norway and France became apparent. This became particularly difficult after Hitler's order of 19 July 1940 to strengthen the Navy and Luftwaffe. Neither Lorenz nor Telefunken had their radar ready for production and so were unable to relieve GEMA in this. Furthermore, their equipment only met the requirements of Flak and was of little use for air warning and coastal surveillance.

Even Flak, which set their hopes for night control of antiaircraft guns on Telefunken's Würzburg and Lorenz's Kurpfalz sets, placed demands on GEMA, as they were under strong compulsion to defend Berlin effectively. Because they could not expect any of the special gun-laying sets until the end of 1940, they wanted to be included in GEMA's distribution.

It is scarcely believable how GEMA met the demands of their customers for the 150 radars delivered by the end of that year, even when one realizes that by then the number of their employees had risen to 2500. One must not forget that, in addition to the uninterrupted production of radar sets, there were extensive commitments for technical support of the equipment in service.

Not until the turn of 1940/41 were there deliveries of complete rotatable bases from AEG's Hennigsdorf plant, and until then GEMA had to fabricate them in their own shops and with the help of diverse subcontractors. During the summer they had delivered and set up seven Freyas with demountable bases for air transport to the Scandinavian coast. For the further protection of that area the Navy needed 15 Seetakt sets with rotatable bases for gun laying by coastal-artillery and antiaircraft batteries. After the occupation of France was complete and the plans for Operation Seelöwe began to take shape, radar had to be set up on the Channel coast, especially at the Dover Straits. In addition to these calls for equipment, Diehl's group, which was experimenting with radar control of night fighters, also required radar, especially sets with rotatable bases for Flum.

The first bases, delivered in 1939 and early 1940, were relatively simple and individually made. Some of them had a wooden-plank support without antenna lifts and had to be traversed by hand. In co-operation with the AEG divisions affiliated with naval, air and ground forces, GEMA developed an all-metal rotatable base on which both Seetakt and Flum sets could be mounted. The new radar generation was so configured that DeTe-I and DeTe-II had interchangeable parts except for the high-frequency elements. Since 1938 GEMA had secured from AEG the services of their special technical branch for the development, fabrication and later for the assembly of electrically driven and remote-controlled bases of various designs for the Navy. AEG excelled in mechanical drive techniques and for that reason GEMA turned production of the rotatable base on a mobile antiaircraft gun mount over to them.

After completing the first mobile bases for old models, which the Navy used to some extent in Denmark and Norway, GEMA began delivering in August 1940 the first new bases for the sets being installed near Calais. These sets, which had a pulse-repetition frequency of 1000 Hz and the old model base, were called Calais-A, FuG(Seetakt)39G(gB). The new 500 Hz model became Calais-B, FuMG(Seetakt)40G(gB). The old model had no lifts for the antenna and an array of only two times 10 dipoles. The new model, delivered after summer 1940, had a hydraulic antenna lift for a 2 m by 6 m array with two times 16 vertical dipoles. By the end of 1940 GEMA had delivered 19 Calais B sets to the Navy.

There were discussions in April 1940 between Diehl and GEMA about his experiments in night-fighter control with a Freya, during which GEMA proposed combining the radar with a UHF radio beacon. The guiding beam of the beacon would be automatically directed toward the target tracked by the radar, and the fighter could then be directed over the radar station and onto the beam of the beacon. The pilot could then receive these signals with a normal part of the aircraft's radio equipment and keep himself directed toward the target. Even automatic control of the plane was considered. Such control devices were in use for automatic bomb release.

GEMA tried experiments designed to make use of their IFF as a means of guiding the pilot. To this end the two lobes of the Freya were to be encoded with the Morse letters A and N so that the flier would know from an audio signal obtained from his IFF whether he was too far to the left or right, depending on whether he heard an A, an N, or a pure tone, the last of which indicated the correct course. The pilot should be able to find the exact direction as soon as he was in the radar beam. This system found strong approval among people at the Air Ministry, but simply because the Navy and Luftwaffe preferred sets that used the maximum-signal direction-finding method, it had to be rejected. The necessary pendulous motion of the antenna caused a swinging beam which would have presented the pilot with confusing information, making it difficult to stay on course.

Diehl tried a Lorenz radio-beacon system intended for aiding aircraft to find their airport in bad weather but without success. GEMA could thank these experiments, and Diehl's insistence on them, for letting them resume work on lobe switching, although they were now able to marshal only a part of their development forces to this end, as so many of their qualified personnel were needed for production.

The g-wave antiaircraft gun-laying radar with horizontal and vertical lobe switching, with which GEMA had hoped to compete with Telefunken and Lorenz, was still on the Köpenick tower. In summer 1940 this set, which had been modified to include the Radattel-Peilung (minimum or zero-signal direction finding) that had used continuous sampling of a half-wave delay line, had been used not only for tests but for directing the fire of nearby Flak batteries during raids. This use had demonstrated to unanimous agreement that this was a vastly better way of using lobe switching than the various methods of comparing the signals of the two lobes visually. GEMA developed this circuit together with AEG as a lobe-switching supplement, P, to the basic set. Its composition is described in the numbered sections that follow.

(1) Motor-driven direction-finding delay-line modules WRg for g-wave and WRf for f-wave sets. Rather than using capacitor sampling, these modules extracted the signal along a half-wavelength line, each end of which was connected in opposite phase to one of the two receiving antenna halves, causing a minimum voltage at the middle of the line. The line was given a ring form and its voltage distribution was continuously probed by a motor-driven contact. The horizontal deflection of a CRT for direction finding was synchronized with the rotation of the contact and the probed voltage formed a continuous picture on the screen of its distribution along the line. A separate contact at the mid-point of the delay line produced a vertical line on the screen. Thus the operator could determine at once from the minimum value of the antenna voltages and the-mid point marking if the target was on or off the antenna axis.

(2) Direction observation module PB determined the exact bearing or elevation of the target. In order to present the distribution of voltage on the ring circuit the PB used a dual-beam CRT. One of the beams was deflected by a saw-tooth voltage that corresponded to the position of the pickup contact on the ring circuit. The saw-tooth generator was triggered by the motor-driven pickup. The beam trace on the CRT thus corresponded to the position of the rotating pickup on the ring circuit as it moved from one end to the other. At the mid-point of the pickup the zero was indicated by a vertical line in the middle of the CRT. The distribution of voltage on the ring circuit changed from the sum of both at one end, to a minimum (the difference voltage) and on to the sum again at the other end. This signal, amplified and rectified, was displayed vertically on the direction CRT so that an accurately located target showed up as a saddle-shaped trace with the minimum at the zero marker at the centre of the screen, if the target was accurately fixed. If the target wandered to one side or the other, the saddle trace moved accordingly. The accuracy attainable with this on the Calais-B was 0.1°. The second beam of the direction CRT allowed the operator to oversee all echo signals for the NB and OB modules, depending on the model of the PB.

Diehl did not use this kind of direction finding, Radattel-Peilung, advantageous for air-warning and fighter guidance. He preferred the old comparison method. The Air Ministry did not adopt the offered easy-to-read meter display method of direction finding, although they contributed an amicable clarification for GEMA's priority claims based on the patent application of 1935. GEMA developed for them a similar direction-finding module or modification, P, as it already existed for using a minimum signal for determining direction. As an appendage for the Freya it consisted of two modules as follows:

(2a) Direction switch with delay line PU. A motor-driven switch alternately connected the two antenna halves to the receiver input cable. Synchronized with this, the output of the direction receiver was switched to display the two antenna signals inverted relative to one another or with a lateral displacement.

(2b) Direction observation module PB. This module had two dual-beam CRTs to make the direction signal visible for comparison. It was similar to the NB module that was part of the main unit N. For direction finding without the need to oversee all returned signals, such as was encountered in directing searchlights, it could be used instead of the NB. A CRT having a vertical time axis served for coarse direction finding. The received signals were directed to the left or right of this axis in synchronization with the antenna halves then connected. One could see from the difference in amplitude of the echoes, if the target lay to the left or right or if it were accurately located. For accurate direction the echoes were examined on the second CRT in order to judge equality with the traces next to one another. The output of the OB module was displayed on the second set of deflection plates of this tube. All time axes of the PB had the same scale and were connected to Messkette OK. The echo being examined was enhanced by enhanced beam brightness.

With these direction supplements GEMA could at the end of 1940 meet all radar requirements. The radars that were equipped with them could, through exchange or augmentation of specific sets, be used for any purpose. Now lobe switching for elevation could be attained by adding the vertical supplement NH, consisting of an antenna, receiver NE and indicator PB. The vertical supplement used the delay-line method of determining the minimum signal.

The Messkette was now dimensioned to its final form of 200 km with a resolution of 100 m, attained with a pulse repetition rate of 500 Hz. For artillery gun-laying there was a special Messkette for 50 km with a resolution of 10 m. This artillery variation had only two range wheels; one revolution of the crank corresponded to 1 km and was coupled to the 50 km wheel. The artillery Messkette used only a single crank.

A whole series of developments was added to the standard series by the end of 1940 despite the infamous prohibition on research and the tensions created by the war.

GEMA equipment took on even greater importance in the war at sea, but there were reports of radars failing to observe echoes owing to frequency misalignment of transmitter and receiver. In order to ensure that radar transmitters as well as receivers functioned properly at all times, GEMA brought out tuning instrument X-100, consisting of an auxiliary receiver, a CRT and a pulse-modulated auxiliary transmitter, which were housed in an aluminium container. **The auxiliary receiver was a super-heterodyne with frequency converter, double IF, demodulator and output. For sharp tuning its bandwidth was reduced to 60 kHz. The demodulated signal was applied to the CRT that obtained its time base from the 500 Hz voltage of the set to be tested or tuned. This brought about the CRT synchronization automatically. The 500 Hz voltage was used to produce pulses for the auxiliary transmitter by overdriving and differentiating. This transmitter was loosely coupled to antenna V/W, as was the auxiliary receiver. The modulator pulse of the auxiliary transmitter was shifted in phase from the transmitter of the set.**

With the help of the built-in CRT the auxiliary receiver was then tuned to the output of the radar transmitter, after which the auxiliary transmitter was tuned to the auxiliary receiver, whose pulse appeared shifted slightly to the left from that of the radar transmitter. With this arrangement the auxiliary transmitter had the same frequency as the radar transmitter and generated with its greatly reduced power an artificial echo on the two indicators that was based on its easily identified position. Now the radar receiver could be tuned to this artificial echo. The radar operator could, using the CRTs of the radar and the X-100, check the alignment of his transmitter and receiver and make corrections as required.

GEMA also provided a power meter LM for both g- and f-wave sets. In an aluminium housing mounted on the reflecting mesh behind the transmitter array were a high-frequency rectifier and a dc amplifier. The high-frequency pulse was picked up with an adjustable, capacitive coupling to a transmitter dipole. For control a galvanometer in the power-supply unit R was placed at the output of the dc amplifier, allowing a continuous judgement of transmitter power to be made.

For continuously observing the frequency of a radar transmitter, in 1940 GEMA provided absorption frequency meters with cavity resonators for the g and f bands. The device was connected with a cable to the set under study and the frequency read directly in MHz from a scale. The indication that the input frequency agreed with the tuned frequency of the instrument came through the small cathode-ray tube that was widely used for tuning radios during those years, called in Britain and America the 'magic eye'.

The Luftwaffe gave high priority to completing the design of the ground receiver to be used with the IFF set (called variously ES, Erstling and FuG25A). The system first consisted of the IFF receiver PE, 'Gemse' (chamois), for Freya with amplifier PV, 'Gemsbock'.

Gemse differed only slightly in its circuitry from the f-wave radar receiver NE. To make operation simple by avoiding tedious tuning, the oscillator frequency was wobbled by a motor so as to ensure reception of signals from IFF sets whose frequencies were not exactly aligned. In typical GEMA form, Gemse was fashioned from cast aluminium. The chassis for the IFF receiver–transmitter Erstling was even more compact.

Gemsbock was developed in 1941, initially considered as a transition device, to determine the direction to the IFF source. Inasmuch as the IFF return was stronger than the reflected radar echo, it made sense to determine its direction with a receiver having a lobe-switching antenna of its own. It was no problem to display the IFF response together with the echo using a dual-beam CRT. The circuitry of the various systems and their cables was expanded so that the IFF return was available to any of the system's dual-beam CRTs not otherwise in use. The time bases were the same for both traces. The pulses that came from an aircraft equipped with IFF appeared on the screen below that of the echo pulse. This allowed the radar operator to see immediately the spatial relationship of the IFF-equipped machine to other targets.

In a series of IFF experiments made in late fall 1940 with a Freya at Jüterbog much greater ranges were attained than with the normal radar. High-flying aircraft were accurately located to the west of Hanover and near Hamburg at ranges of 250 km. After this it was decided that all Freyas would be equipped with IFF systems capable of determining direction.

In fall 1940 GEMA had to undertake a special design for the Luftwaffe that had resulted from work in Diehl's command. In order to direct searchlights while still waiting for the undelivered Telefunken and Lorenz radar, Diehl had mounted dipoles with Freya wavelength to the left and right of a searchlight. To this antenna he connected the Freya lobe-switching electronics and in this way he could use the reflected waves from a target illuminated by a nearby radar to point the searchlight in elevation or bearing, depending on which pair of dipoles was connected.

GEMA developed two vertically polarized antennas for the 200 cm searchlight Type 40 that could be mounted to the searchlight housings. An arm was attached to the searchlight on which the receiver and indicator were fixed at the height of the searchlight operator, who pointed the searchlight according to the indications of the CRT until the target was picked up by the light beam. This passive radar received the name 'DeTe-Schein' and was built for f and g bands. The Luftwaffe and Navy ordered a total of 70, but the only deliveries were 17 to the Navy in spring 1941. It is not known whether this device saw action. Telefunken had in the meantime begun deliveries of the Würzburg, which superseded the GEMA passive system.

In August 1940 GEMA explained to the Air Ministry their progress in devising an airborne radar, called 'Klein-DeTe-Gerät'. The Ministry had questioned Erbslöh earlier about the possibilities for such a radar, and he could point to preliminary work along those lines, work that had come about during the IFF development. After six months there was a prototype having a weight of 20 kg and a volume of 500 litres. It had the typical hallmark of a DeTe set. In order to use the essential parts of the IFF, they used the same frequency as this device. At the beginning of 1941 three of these sets with a variety of antennas were delivered to the Luftwaffe test station at Werneuchen. In comparison with equipment furnished by the competition, the GEMA sets were too heavy and could only be mounted in large aircraft. Only at the turn of 1941/42 did the set reach a service trial from which came the sea-search radar 'Rostock'.

With the introduction of the Telefunken and Lorenz Flak radar on frequency 560 MHz there arose the problem that GEMA's IFF would not respond to them, as it worked on the f band. Originally GEMA had intended to design their IFF for use with the g band of Seetakt or Navy Flak radar, but the discussions with the Luftwaffe and the Navy had led to the adoption of the f band for direct interrogation, and GEMA had incorporated this frequency into the ground and airborne IFF elements.

In order to adapt the new Flak radars to their IFF on 2 m wavelength, GEMA provided a supplement to the Flak radar in the form of a special interrogating transmitter Q with the cover name 'Kuh' (cow). It consisted of unit QU with pulse generator and output stage and power supply QN, which furnished the voltages for the pulse generator and output stage. In the first models the ground unit for identification, Gemse, and the unit for interrogation, Kuh, were connected to separate antennas, but later GEMA developed a duplex antenna for both transmission and reception, which relied on research done in making radars with common antennas.

Under the force of circumstances—the RAF night attacks were becoming steadily more unpleasant—the effective range of the Freyas used for air warning absolutely had to be extended. Since summer 1940 there had stood at Jüterbog a second 30 m high climbable mast mounted on a powerful rotatable bearing, which was fixed to a concrete foundation. At the top was a second bearing from which guy wires steadied the assembly. This mast was an essential part of DeTe-Weit (far) or DeTe-W. From this letter came the cover name 'Wassermann', the radar that ensued from these experiments.

Following the rule that 'the best high-frequency amplifier is the antenna', GEMA was able to increase the range of the Freya in these experiments by 30%. This path was chosen because at the time, increasing the size of the antenna seemed simpler than increasing transmitter power or receiver sensitivity.

In connection with these range experiments, which were carried out with an IFF as echo enhancer on the Brocken, 170 km distant, the Luftwaffe placed at GEMA's disposal a 45 m climbable mast on the King of the Hessian Mountains, the 750 m high Meissner. This mast, which was on the grounds of a Luftwaffe signals unit, held a GEMA IFF set to serve as echo simulation. An arm on the mast allowed an 8 m by 8 m frame with reflecting foils to be hoisted to simulate an aircraft. The location on the Meissner was selected because the land lying between it and Jüterbog was favourable for f-band propagation. The distance attained satisfied the hopes of what a Freya with a sixfold area of antenna could bring about; the mast at Jüterbog could hold no more arrays.

The expectations were confirmed in experiments in early 1941. After simulations on the Meissner there were experiments on a Ju-52 and a He-111 that flew back and forth at high altitude and that were tracked at 280 km with a pulse power up to 20 kW.

The Luftwaffe showed much interest in this new, stationary Wassermann and pressed for completion of its development. For the mast GEMA chose a capable subcontractor. For their first set they designed, together with the firm of G Seibert, a relatively light, triangular-frame mast that could be transported on trucks and that held six Freya antennas. The foot of the mast was made from a cylinder that was turned through 360° by AEG gears. The console and electronics were set on a platform around the cylinder, all within a metal-walled cabin in which the operators worked. The entire structure was 37 m high and rested on a strong foundation. Guy wires secured the mast from the bearing at the top.

Seibert constructed the mast, and GEMA the cabin, its contents and the antennas. The basis for the construction of the cabin came from the demountable equipment built for the Luftwaffe for the invasion of Norway. Because of the need for speed at that time, the cabin could not be made in a round or multi-sided form. The result came fast: Wassermann-L, the letter L standing for the light, aluminium-frame mast. Construction techniques learned from this work led later to Freya-LZ (Lufttransport-Zerlegbar, air transportable, described later) that GEMA manufactured together with Luftschiffbau Zeppelin GmbH, Friedrichshafen.

On the basis of Wassermann-L GEMA began in mid-1941 the development of a rotatable mast on which twelve Freya antennas could be hung at the initiation of the Luftwaffe. It was, of course, accepted from the start that such a heavy antenna support would have to be constructed over a concrete bunker as a stationary emplacement. This proposal, which differed from Wassermann-L especially in having double the antenna area, received the cover name Wassermann-S (schwer, heavy).

Wassermann-S used a mast like the smaller 'Zerstörersäule' (destroyer column), a large-diameter cylinder of the same kind on which the Seetakt sets were mounted in light warships, such as destroyers, that did not have the large optical rangefinders on which the Seetakts were normally mounted. The column established the antenna's bearing and was remote controlled. The continued attention paid by Allied aircraft to radar stations led GEMA to build bunkered control rooms on which the rotatable column sat.

For Wassermann-S the mast was placed on a bunker with a deep foundation. The mast was made of several steel sections and was 47 m high. A geared ring at the base allowed 360° rotation about the vertical axis at two speeds. Owing to the large cylindrical column, British intelligence referred to the Wassermann as 'Chimney', a term that remained despite the varied constructions of the tower.

In their 1939 experiments with lobe switching GEMA had tried various means for slewing the radiation lobe using electrical methods. The basis was that the signal received by each dipole in an array had its own transit time to reach the receiver. In order to determine the direction from which the wave arrived one must control the relative phases of the individual dipoles. For the Wassermann sets there was a need to slew the lobe in elevation in order to determine the altitude of the target. The first method tried used capacitative compensation for slewing the radiation lobe and for connecting the dipoles of the array to the receiver. The output of the receiver was observed on a CRT screen with a voltage proportional to elevation angle on the other co-ordinate. This arrangement was simply not satisfactory in its mechanical usefulness and was replaced by a system in which individual arrays were connected, according to their location on the mast, by variable lengths of transmission line to the receiver. Schultes invented this compensation system and called it his 'Wellenschieber' (wave shifter). It was later incorporated in the final design by Siemens as Wassermann-M.

A problem for GEMA, one that was mounting in its seriousness, was the attitude of the Navy and the Luftwaffe about training military operating and maintenance personnel for equipment that was continually becoming more complicated. The Luftwaffe had made itself more or less independent of any schooling and training by GEMA since the middle of 1941. They organized within their signals units their own radar service with minimal interaction from GEMA. In Navy signals units there was still no radar career path. The task of training lay with the NVK (the successor of the NVA) in Kiel, with the rangefinder school at Sassnitz and especially at GEMA, there under the supervision of Lieutenant von Willisen. The NVK involved themselves primarily with the training and guidance of operating personnel aboard warships. Von Willisen with a number of the company staff supervised untiringly the construction of radar stations along the coast from the North Cape to the Spanish border, and profusely along the Channel. As a man who always wanted to be on the spot and with his wide experience in radar and in working with naval command, he represented GEMA to great advantage, and their leading position as manufacturers of radio-location equipment owed much to him. It was valuable use of his time but detracted from his services in expanding training.

Only at the end of 1941 was a career position opened for radar within the Navy. This happened at a time when a fight between warships could hardly be imagined without radar. The first radar section of naval signals was organized at De Haan near Ostende in December 1941 and undertook regular schooling for radar operators under the supervision of von Willisen and GEMA personnel.

After General Ernst Udet, master of air ordnance, committed suicide from despair at the Luftwaffe development failures, the planning section of the Air Ministry decided to form a research group, the leadership of which was turned over to the engineering departments of various arms manufacturers. The leadership for radar was given to Leo Brandt, at the time development chief at Telefunken.

In 1941 the 3000 employees of GEMA delivered to the Navy and Luftwaffe radar and sonar equipment worth more than 35 000 000 RM ($8300 000). The equipment total was 250 sonars, 122 Freyas and 208 Seetakts in various forms. There were also production deliveries by GEMA subcontractors, especially AEG. The special technical branch of AEG worked well with GEMA and had developed the indicator with dual-beam CRT. There were some initial problems because the units did not meet GEMA's tough mechanical test requirements, which required that they endure vibrations of 5–20 Hz and 3.5 mm amplitude successfully for three hours. After a redesign of the dual-beam CRT, AEG was then able to meet GEMA's standards.

GEMA fell into severe difficulties at the end of 1940 in delivering rotatable bases. The Navy had to equip the coasts of the occupied territories with Seetakt sets for coastal artillery and had to prepare defences for them using their own Flak radar against increasing air attack. Equally pressing were the demands of the Luftwaffe, which had had to cope since mid-1940 with strategic bombing by the RAF. During the night of 7/8 June the first bombs fell on Berlin. This raid, laughable in its effect at the time, had escalated by the end of August, as squadrons of bombers dropped their deadly cargo at night and departed with meagre losses.

Diehl's experimental method of dunkler Nachtjagd (dark night fighting) was urgently pushed to combat the night bombing. It was an elaboration of heller Nachtjagd (illuminated night fighting), which depended on searchlights and was thus strongly affected by the weather. Because the night-fighter controllers needed Freyas, there was an immediate demand for more of these sets in the regions across which the RAF bombers passed. In order to discover and locate quickly the approach of enemy aircraft at night or during bad weather, the organization of an effective air-warning service was required, and it needed long-range radar.

The collaborative work with AEG for the manufacture of rotatable bases was not impressive by the end of 1940, with substantial delays in delivery. Even the production of dynamotors for the bases dragged along at AEG. By mustering their entire shop and assembly capacity GEMA was able to meet the demands of their troublesome customers without serious retribution. Erbslöh found himself bearing the brunt of growing criticism of GEMA, especially from the Navy, who saw production for the Luftwaffe as detrimental to their interests. Erbslöh understood how to calm those who were above the ones who were complaining loudly and was able to fend off direct control by the Reich Fiscal Office until the end of 1941.

I have described in some detail GEMA's inventive exploits to provide a historical appreciation of their contribution to radar. In this regard they developed in 1940/41 much special measurement and test equipment for the maintenance of their equipment in the field.

Even considering that much of the development had begun before 1940, the accomplishments of the laboratories under Schultes and Brandt are remarkable in 1941. Many of the steps that they had to take to reach a useful radar system were in a region filled with technical unknowns, steps which were formally forbidden in Germany after 1939 by the infamous 'stop research' order.

It is understandable that GEMA's management was proud of their firm and its achievements at the end of 1941, but they did not fail to recognize the envy and ill will that now surrounded them.

CHAPTER 17

TURBULENT TIMES

Not until 1941 did a competitor to GEMA arise that had to be taken seriously, both in sonar as well as radar.

Lorenz had in 1935, as has been noted, a contractual agreement with GEMA for production in which Erbslöh and von Willisen had provided licensing to Lorenz of all their radar inventions, developments and constructions. Lorenz thus had access to GEMA's patents and current work until Naval Command required the association of the two firms be terminated, but it was not possible to prohibit Lorenz, with or without the advantages accrued during the time of their co-operation with GEMA, from entering the field of radar on their own.

Telefunken was well aware of GEMA's successes in radio location. In early 1934 Dr Wilhelm Runge, a scientist in Telefunken's receiver laboratory, had reacted in a most negative manner to Kühnhold's suggestion of a collaboration in the new field of the reflection of high-frequency waves. Telefunken had begun, on the basis of their decimetre-wave research to which Hollmann had contributed for a time, the development of communications equipment working on 50 cm. In 1935 Runge became the leader of the decimetre-wave laboratory in which preliminary experiments were conducted on a transmitter and receiver for the directed-beam communication links, Telefunken's 'Olympia System'. At that time Runge demonstrated to himself with a special transmitter, which produced 5 W on 50 cm from the TS1 triode, reflections from a Ju-52 at an altitude of 5000 m that flew over his apparatus. These results yielded little appreciation among the company management, so more or less secretly he conducted radar experiments along with those aimed at directed beams.

In fall 1938 Telefunken began the development of the Würzburg on the basis of Runge's preliminary work. Runge and his co-workers established the new concepts in a set called Darmstadt that matured into the Würzburg, which in time evolved into the gun-laying sets Mainz and Mannheim. In mid-1939 the first model A3 underwent tests. With it, a simple one-man-operated set, the bearing and elevation of an aircraft was determined to an accuracy of 1.5°. The accuracy of its range determination was 100 m at 30 km. This radar was quickly introduced into the Luftwaffe, but Telefunken did not drive GEMA from the field of antiaircraft radar with it.

In summer 1940 GEMA began to feel Telefunken's superiority in production as well. The delivery of Würzburgs came with startling speed, which to GEMA's relief displaced Lorenz from the field of Flak radar through their better suitability, leaving GEMA with only one competitor, although to be sure a very strong one. Once, however, they were able to register a small triumph while employing a Freya in combination with a Würzburg in the air defence of Berlin. During a long attack in December 1940—it lasted almost the entire night of the 29th— the Würzburg sets being used by air defence failed, and only GEMA's Freya returned bomber positions. The reason for the failures was the extreme cold that affected things in two ways. The crew was simply not able to function in the extreme cold, as the Würzburg had no heated cabin, in contrast to the Freya next to them. In addition the Würzburg failed because of frost and condensed water. The mechanical controls froze and the water produced caused repeated short-circuits. At that time Flak trained their operators in a 14 day course at the artillery school, and they had had little experience with radar. Later the instruction period was extended to three months.

Telefunken developed the Würzburg into a precise gun-laying set for Flak by introducing lobe switching in the form of a dipole rotating off axis at the focus of the paraboloid. The Würzburg was originally designed to follow close-in aircraft and could not contest Freya's range, the consequence of its 53 cm wavelength. The narrow beam attained with the 3 m paraboloid, which allowed it to track accurately targets first located by a Freya, made it a requisite for fighter control. In mid-1941 Telefunken and Luftschiffbau Zeppelin equipped the set with a 7.5 m dish, a cabin and the electronics of Würzburg D to form the Würzburg-Riese (giant), which was capable of tracking targets from 50 to 70 km. The first of these was mounted on top of the Flak bunker at the Berlin Zoo.

GEMA countered the challenge of the Riese with their first panoramic radar, complete with plan position indicator (PPI). Erbslöh had suggested a device with a rotating directional antenna that explored the horizontal region about the station with high-frequency pulses in a patent registered in 1936. The receiver antenna, at the time planned to be separate, was synchronized with the angular motion of the transmitter antenna and displayed the echoes observed by modulating the brightness of a CRT beam. The deflection of the beam was accomplished with a magnetic coil positioned on the neck of the tube that rotated mechanically with the antennas. The CRT beam traced a path from the centre of the tube to the edge of the screen with its direction synchronized with the antennas. Since only echoes brightened the trace, the screen took on the appearance of a map showing the fixed and moving reflections.

In January 1939 Erbslöh had explained to the officers of the Luftwaffe and Flak at Lynow the possibilities of cartographic representation on a CRT.

At that time a laboratory version of such an indicator with a mechanically coupled deflection for displaying a sector scanned by a radar was available. Although Erbslöh did not notice great interest in this method of display, he permitted the research to continue. Out of experiments with a rotating Freya, at first on the Köpenick tower and later at Jüterbog, came the predecessor of the panoramic radar 'DeTe-Panorama' that later became 'Jagdschloss'. In 1941, after a demonstration in Berlin, GEMA received from the Luftwaffe an order to deliver a panoramic radar, with the Zoo bunker or some nearby location proposed. When GEMA learned that Telefunken was going to build a large radar on the Zoo bunker, they could no longer consider using that location. They also learned that the Telefunken set was expected to reach the range of Freya and that, owing to its greater accuracy, could possibly drive them from the field. This was the background that caused Erbslöh to give high priority to the construction of the panoramic set.

At the end of 1941 GEMA had Siemens build the drive mechanism for continuously rotating a horizontal antenna mounting on a platform 20 m high on a masonry tower in Tremmen, 35 km west of Berlin. This mounting was a 20 m length of the triangular-frame mast of Wassermann-L that was fixed at the centre to the drive mechanism. The drive rotated at 6 rev/min and the frame carried 18 full-wave dipoles with reflectors. This antenna served as transmitter and receiver for the Freya electronics connected to it. Separation of the transmission from reception was accomplished with a duplexing circuit. For the first indicator GEMA had used a magnetically deflected CRT of 200 mm diameter from Leybold/von Ardenne Oszillographen GmbH. For the indicator at Tremmen they developed with Fernseh AG a PPI from a magnetically deflected television tube of 380 mm diameter. In this set a selsyn pair caused the motion of the deflection coils. The selsyn transmitter was mechanically connected to the drive mechanism and electrically connected to the selsyn receiver by a cable, thus turning the deflection coils. After notable troubles with the dimensioning of the duplex circuit within the antenna cabling, which was covered in order to protect it from the weather, the set at Tremmen went into operation in March 1942. It found strong approval from the Luftwaffe. In summer 1942 the PPI was provisionally transmitted through a Post Office cable to the Zoo bunker. The synchronization of the pulse repetition frequency and the rotation of the antenna was included in the modulation of the carrier that transmitted the reflection signals.

By the turn of 1941/42 there were more than 250 Seetakt and Freya sets on fixed and mobile mountings for sea and air surveillance in service with the Navy and Luftwaffe, and the first Wassermann sets would soon be in service for distant air warning. The transportable Wassermann-L sets were sent preferentially to the Mediterranean, as they were relatively easy to set up. The Wassermann-S sets required a comprehensive foundation to be constructed. The first training set was installed on the west coast of Denmark where it served to track the RAF bombers headed toward the Reich by paths avoiding the direct way over Holland, choosing instead the North Sea and Denmark. At the Channel one received from the Freyas a satisfactory picture of the air space for both day and night fighters, but stations in Scandinavia were out of range for those sets.

The Navy again went its own way. At the start of 1941 GEMA, despite their heavy commitments, had to design in collaboration with the NVK a special set on the g band for long distance air and sea warning, designated 'Mammut-Gustav', FuMG41G(gA). A stable electronics cabin with a grid-like roof construction for attaching six antennas of the Calais type on a concrete foundation was placed on rails that allowed the entire assembly to rotate about its vertical axis. It was equipped as a pure search and warning station with DeTe-I electronics and only determined the direction by maximum signal. The range was about the same as the Würzburg-Riese, and because the Navy preferentially ordered these sets from Telefunken, only four Mammut-Gustav sets were manufactured.

While the Mammut-Gustav was being built, the Navy ordered a radar with phased-array slewing on the newly introduced 1.5 m wavelength band c, from 182 to 215 MHz. It was codenamed 'Mammut-Cäsar', FuMG41G(cF). It resembled Mammut-Gustav except that the antenna was fixed. This set was intended for sea surveillance in the defence of the Atlantic Wall. On top of a bunker ensuring safety for the crew and the electronics stood three 20 m high masts to which were affixed six antenna groups looking seaward. As a consequence of the longer wavelength the size of the antennas had almost doubled over those of Seetakt. Phase compensators allowed the radiation pattern to be deflected ±50° from the central axis. A total of 16 sets were delivered, not enough to cover the entire Atlantic Wall without gaps.

Primarily for air-warning service GEMA built both for the Navy and Luftwaffe the f-band 'Mammut-Friedrich', FuMG(Flum)41G(fF). The set was also mounted on a strong bunker. It served well along the coast from Norway to France. On or near the bunker stood four 12 m high frame masts in a row that were connected about halfway up with a catwalk construction and from which were hung an upper and lower row of four complete Freya antennas; in some versions the antennas were hung on both sides. Originally half of the antennas were for transmission, the other half for reception, but after a duplexer with protective gas spark-gaps was finished all dipoles were used in common mode. A helical line compensation network altered the lengths of the transmission lines feeding the antennas and allowed deflection of ±50° from the central axis. The large vertical extent of the beam allowed targets to be tracked until they passed over the station, and those stations having antennas on the reverse side could continue tracking inland. With this double-sided arrangement aircraft could be observed within 200° and for 300 km. Owing to their wall-like appearance, British intelligence named the Mammut sets 'Hoarding' (or Billboard in American English).

The training station for Mammut-Friedrich was on the island of Rømø off the Danish west coast where it watched for aircraft coming from the English Midlands toward the Helgoländer Bucht. This set began the first regular long-range surveillance off the Danish coast. It served for training and hence was always outfitted with the latest technical alterations. Fate held in store something special for this radar after Germany's capitulation.

On 4 May 1945 the last Commander of the German Navy, Admiral von Friede-burg, negotiated the surrender of the German Wehrmacht at Field Marshal Mont-gomery's headquarters in the vicinity of Lüneburg. Among other things imposed on the German delegation was that all equipment relating to air surveillance and control, especially those in Denmark, were to be surrendered intact. Technical experts and scientists of the RAF prepared an investigation to secure an under-standing not only of the German equipment but also of how it functioned in the air control system. The thoroughness went so far as to gather together former operating personnel who had already scattered. Starting on 25 June 1945 they were required to take their places at their radars again and 'play war' in whatever manner the observing victors pleased. After this scarcely known action, called 'Operation Post Mortem', all equipment was disassembled and distributed among the Allies. The station on the Island of Rømø remained in operation the longest because it was the most powerful GEMA set, being capable of tracking high-flying aircraft at 400 km.

Starting in 1942 friction began to mount with companies that were certainly not competitors of GEMA but worked for them.

Already in 1940 AEG had designed a so-called unified generator for sonar with the help of the Navy, although GEMA with the knowledge of the Navy had developed a new model. The AEG sound-pulse generator consisted of a self-exciting push–pull circuit with two AS1000 output tubes. It was smaller than the GEMA generator, weighed less and was less expensive, and GEMA could not prevent it being installed in newly constructed U-boats after 1941. Delivery was direct to the shipyards, but the overall responsibility of GEMA for sonar was not yet affected. This inclusion of AEG in the distribution of sonar parts was not a surprise for GEMA. Naval Ordnance had indicated on several occasions that they planned to increase their dependence on AEG for location equipment. GEMA was able to check the placement of orders for complete radar sets with other firms by refusing any responsibility for them. They drew additional subcontractors into their production and in this way removed any need for the Navy to place such orders.

The extension of the war, especially the invasion of Russia and the declaration of war on America, increased the requirements of the Navy and Luftwaffe to such a degree that von Willisen and Erbslöh could hardly see at the turn of 1941/42 how GEMA could maintain their position as the primary radar supplier, and the Navy insisted in an annoying manner on exerting their influence. It became increasingly difficult to avoid becoming an armaments industry controlled by government fiscal officers. AEG had attempted several times, directly or through the Navy, to change their relationship from a subcontractor to a partner. Lorenz was just as eager as AEG for a close association and wanted to revive the contract of 1935. Finally Telefunken and Siemens saw profits in producing GEMA equipment. Erbslöh and von Willisen had nothing against working with these companies, indeed the war situation left them no other choice. But they did object to turning over to competing firms their designs and experience. Looking forward to the time after the war, they knew that radar would have important civilian uses and that what was now co-operation would then become competition. In adroit negotiations GEMA's managers succeeded in imposing an agreement about postwar work as a preliminary condition for any additional close collaboration with AEG and Telefunken. Both companies told GEMA that they would be ready to conclude an agreement after the war that allowed further work to proceed on a basis that would preserve existing patent rights. On the strength of this GEMA agreed to allow these companies to manufacture copies of their equipment and to modify and extend developments of them. GEMA insisted that these companies be given only those documents needed for production and refused them any examination of patents.

On the advice of the Air Ministry Lorenz decided against entering into such an agreement. They alluded to their earlier relationship and indicated the wish of the Air Ministry that the two firms develop together new radar for ground and airborne use by the Luftwaffe. At the start of 1942 Lorenz offered GEMA a very close collaboration wherein they would assume responsibility for 50% of GEMA's manufacture. Lorenz had realized that their own development of radar lagged far behind GEMA. In the competition for a Flak radar Telefunken had cut Lorenz out and gained a gigantic lead in the process, so a marriage with GEMA would put them in a strong position relative to Telefunken. Lorenz was the principal supplier of communications equipment for the Luftwaffe, who hoped to secure advantages in delivery from a Lorenz–GEMA marriage. It was widely known that the Navy, on the basis of the loan they had provided GEMA in 1937 for the expansion of production, insisted on having priority in delivery.

Lorenz's proposal for an alliance was rejected by the two GEMA companions, but Erbslöh was able to convince them, drawing on the old association, that even without an alliance a close connection with Lorenz had advantages for Lorenz as well as for the Wehrmacht Fiscal Office. In April 1942 GEMA and Lorenz drew up a contract that regulated their co-operation in the construction of radar equipment and, depending on the wishes of the Wehrmacht, that allowed common development work. It was agreed that, regardless which of the two had received the order, one or the other would oversee the job. Each firm retained the right to apply for its own patents; the other firm would receive a simple non-transferable license. Patents on inventions made in common would be applied for together. The agreement covered all development contracts that the two partners granted in common or that one of them granted the other.

The closest relationship for technical collaboration was with Siemens & Halske. Already in mid-1942 the GEMA laboratories worked with those of Siemens on matters that were not covered by a contract until 1944, the outcome of long negotiations that extended its application retroactively to 1 October 1942. The purpose of the contract, which was arranged in the interests of the Luftwaffe, was to join the two companies for the duration of the war in developing radar equipment to the highest possible degree. GEMA came to an understanding with Siemens that was to lead to a friendly relationship after the war. The two firms agreed that, both in war and peace, the material objects appertaining to the contract would of course be developed, fabricated and distributed in free competition. The material aspects of the contract covered not only radar equipment but also the vacuum tubes such equipment required. It was natural that GEMA was to play the leading role in such associations with its long experience in the development and manufacture of sonar and radar. Now that they were, at least in the field of radar, a renowned firm, they experienced an uncommon rise in their worth.

Owing to their concern for Kühnhold's interests and the Navy's secrecy requirements, GEMA avoided taking part in many technical meetings relative to their work. This even included participation in Arbeitsgemeinschaft Rotterdam (Rotterdam Committee), which met to consider the consequences and the best response to the discovery of the resonance magnetron that had been found in the ruins of a British bomber in February 1943, although Kühnhold and GEMA had been informed of the discovery. Kühnhold was very much interested in this development and had continued research in centimetre waves at his own laboratory with Dr Anton Röhrl. GEMA did not think microwaves would have much impact on the long-range air-warning radars that they took to be their principal responsibility, and the application to naval radar was not immediately clear.

One notes only as an aside that the Navy was not pleased at the evolution of their progeny. It had not been the intention of Erbslöh to form a particularly strong relationship with the leadership of the Luftwaffe, but as it came to pass that inasmuch as von Willisen, an officer of the Navy, naturally represented the company interests there, so it fell to Erbslöh to deal with the other service. Since von Willisen was seldom on duty in Berlin, there existed no close contact between the naval agencies there and in Kiel. Indeed, von Willisen had his office with Naval Command West in Paris. He noticed some time later that there was a growing unrest about his company in Kiel and Berlin.

By the end of 1942 GEMA had manufactured almost 1500 of their well proven sonars. Already at the start of the war at sea it had shown itself to be necessary to have one or two of these sets in master–slave operation on surface ships. As all attempts to synchronize the motor-driven sweep mirrors on the indicator failed, the researchers remembered the early half-silvered mirror displays in which the mirror was moved by an electrically wound clockwork. This old device offered a method of accurate electrical starting of various mirrors in different locations. The result was the GH, another contrivance of electrical releases and mechanical unwinding. With the GH as master one could mount any number of slave units. By July 1941 GEMA had delivered 30 of the new devices. After a preliminary series of 10 sets they delivered 40 slave units per month by December.

From mid-1941 U-boats were no longer equipped with sonar for tactical and space reasons. Sets that had been delivered for that purpose were diverted to equip surface vessels. In order for U-boats to locate enemy ships at night or in bad weather, the Navy ordered radar. In September 1941, therefore, began the series production of the DeTe-U, which had been tested much earlier, the FuMG41G(gU) which was later called FuMO29, and which had antennas fixed to the forward side of the conning tower. The radiation pattern could be slewed ±10° relative to the forward direction with antenna-phase-shifter VK/WKu. The set was composed of a compact block from modules R, T, Nu, Zu and X. Modules R, T and X were of normal construction. Nu was a receiver specially built for the U-boats and incorporated the indicator OB, used for survey. Module Zu was specially designed for this case. It had the audio oscillator and had as a replacement for the heavy Messkette, which had been eliminated, a calibrated CRT screen. The calibration extended to 30 km and satisfied the naval requirement of 100 m resolution. Direction was obtained by maximizing the reflected signal.

By 1942 there had been 75 sets of this design delivered for standing orders. After that 360 improved models, FuMG42(gU) or FuMO30, were delivered. The later set was distinguished from its predecessor in having a steerable antenna on the mast. Especially for this antenna, which could be mechanically rotated 360° and which was stowed in a pocket-like construction on the side of the conning tower, GEMA developed pressure-tight antenna matching units.

The U-boat radar is mentioned because modules Nu and Zu differed from the usual Seetakt sets. Seetakt sets appeared on warships in over 30 different models. Single sets were always the same and for a given application were put together from the standard GEMA programme. The antenna was, however, often manufactured or assembled with consideration for local conditions, work often done by the Navy. Quite important were the 52 antennas GEMA delivered for shipboard Flak for large warships, which could determine elevation or bearing. For installation at the shipyard GEMA's duties as a rule were for supervision, adjustment and removal of radar sets.

The employment of GEMA's increasingly strong research departments had by autumn 1942 grown to such an extent that the managers had to decide to undertake new projects through the use of subcontractors.

The collaboration with Siemens had in the course of time developed itself in so gratifying a manner that AEG and Lorenz were called on less frequently. Before the start of the war AEG had been tied to GEMA's deliveries and had subsequently often attempted to go around GEMA in obtaining relevant contracts, on the basis of their goodwill with the Navy. This generated a wariness with GEMA that impeded their working relationship. It was known that GEMA was engaged in preliminary negotiations with Siemens concerning a general contract about co-operation in radar that would expedite Siemens's entrance into radar by giving them insight into the advanced technical state of GEMA's designs. GEMA hoped for compensation from Siemens in securing personnel and support for new development. The acceptance by Siemens of the production of certain products would give GEMA needed relief. In the area of remote control, especially the transmission of data electrically, GEMA had depended on Siemens for a long time.

In October 1942 GEMA signed an agreement with Siemens to take over the Wassermann project. It turned out that the development was not yet complete, although 20 sets had been ordered. During the tests a number of matters had presented themselves that had to be incorporated in the final design.

Since March 1942 there had been an imperative requirement that all radars had to be outfitted for operation at variable frequencies in order to be able to work through jamming, which had been encountered with Seetakt equipment. On the night of 27/28 February 1942 a British commando unit had raided the Luftwaffe radar station at Bruneval on the French coast and obtained the important components of a Würzburg from which they would have obviously learned that the transmitter and receiver worked on a single frequency. GEMA had already altered their g-band equipment to allow tuning, so they set about doing the same for f band. The receivers were tunable over a certain range and did not require alteration. In September 1942 GEMA shifted the wavelengths of the pairs of f sets at Douvres, Cap de la Hague, Guernsey, Toqueville and Cap d' Antifers to ranges of 2.32–2.48 m by altering the plate loop, which determined the transmitting frequency. The resulting mismatch of the narrow bandwidth antennas impaired performance markedly.

Because the jamming of the Wassermann sets would have a serious effect on the air-warning system, it was necessary that these sets had to work on other frequencies without loss of distance capability, and this required that the antennas be optimally matched for all frequencies; they had to be provided with broad-band antennas for their entire operating range.

These urgent but completely new problems came at a time for Schultes and his HF laboratory when he was obliged to increase the powers of both g- and f-band transmitters. Strenuous efforts had brought the g-wave transmitter to an impressive 8 kW with the grid-modulated TS6. Success was greater for f waves using the TS4 and later the TS41. With a new control unit TS and with improved circuitry, the TU's power was increased from 10 to 30 kW. It was a level attained with grid modulation that could not be exceeded. The TS6 was to be replaced by the TS60, then being developed, which was to work with 400 W heater power and 20 kV anode potential. For f-wave transmitters the development had begun on TS100 with heater power of 1 kW and 60 kV anode potential. The urgent demands for tube production so oppressed that department that little time remained for these improvements. Tube research people even had to help on the production lines for the TS1 and TS1A, which were in critical demand at the front.

In 1938 Group Leader Tigler in Brandt's low-frequency laboratory had produced a design for a U-boat sonar pulse generator that used a small, light inverter that used the main batteries of the vessel. This required the generation of an alternating current of 15 kHz with 3 kW from direct currents of 110–150 V. It was a saw-tooth voltage attained through the discharge of a capacitor into a chamber that had a short de-ionization time. This astounding invention, the development of which was seriously checked by the termination in the installation of active sonar on U-boats in 1941, did, however, give GEMA valuable experience in design and application of spark-controlled mercury rectifiers. The solution of these problems opened the door to GEMA's techniques for delay-line spark discharging of the modulators of high-power radar transmitters.

Since the size of radar antenna arrays aboard ships could not be arbitrarily increased, only an increase in transmitter power remained to meet the demands of the war at sea. The necessity for increasing the effective range was particularly strong for U-boat sets, which could only detect targets up to 10 km, the consequence of the low position of the antenna above the surface of the sea. In order to help as quickly as possible, GEMA developed a low-power anode modulator for the TS6. To generate 18 kV pulses a capacitor was charged to 8 kV and discharged through a spark-controlled mercury rectifier into the primary of a pulse transformer. The pulse so produced at the secondary functioned as the modulated anode voltage for the transmitter. With this set, of which 12 laboratory models were built, the output pulse power was increased to 20 kW, but in practical trials there were frequent voltage breakdowns in output tubes and circuitry, so rapid deployment was not considered, as the risk did not compensate for the small improvement in power; 50–100 kW was actually needed.

The development of anode modulated transmitters was interrupted in autumn 1942 in favour of the conversion to variable frequency operation. Fear of the start of a radar war conspicuous with massive enemy jamming of current radar brought about action under the cover name 'Wismar', whose goal was to make all current radars tunable over some range of frequencies as soon as possible. GEMA was grievously affected, especially in their f-wave sets. The easiest part was the alteration of the transmitters; the receivers were from the beginning tunable over a range acceptable for Wismar. Much more difficult was the development of broadband antennas for equipment based on the Freya system, with and without phased-array slewing. This hit GEMA extremely hard because of the Luftwaffe's merciless insistence on the completion of designs for the panorama-radar Jagdschloss and the Freya with demountable base and antenna assembly.

Erbslöh utilized this situation cleverly to protect himself and his company from possible reprisals by Albert Speer's Office of Weapons and Munitions. In the confusion of events he was able to introduce Siemens as being responsible for compliance with the Wismar project. Although the fulfilment of the contract with Siemens, which now had to be concluded on the basis of a general revision of all contracts with the Air Ministry, was not complete, GEMA transferred the entire Wassermann project to Siemens as was the modification for Wismar. Dr Karl Rottgardt was the chairman of the Sonderkommision für Funkmesstechnik (special commission for radar) as well as Director of Telefunken and did not think GEMA capable of accomplishing these tasks; his support of Erbslöh's decision was very helpful. His help proved useful again at the beginning of 1943 when he supported the transfer of the GEMA–Siemens panorama radar project to Siemens.

The continually improving relationship between GEMA and Siemens was underlined when it was agreed that Schultes terminate his work at GEMA and place himself at the disposal of Siemens. GEMA replaced him with Dr

Kober. Schultes's service meant an extraordinary support for Siemens in entering GEMA's radar field. It eased the fears of both companies that Speer's ministry expressed approval of these arrangements.

Not so cordial, though expected, was the reaction of the Navy about the most recent behaviour of 'their' GEMA. They saw themselves through these measures so brusquely dismissed to the apparent advantage of the Luftwaffe that they withdrew their fondness for GEMA, even for underwater sound location work.

Naval Ordnance told GEMA at the end of 1942 that they were transferring the overall control for sonar to Elac in Kiel, which would direct all future development in this field. Naval Ordnance based this decision by indicating that only a firm dealing in acoustics was competent to hold such responsibilities, reasoning that acoustics was basic, whereas electronics was peripheral. Naval Ordnance limited themselves by ruling that during the war the developmental and production capacity of GEMA and their organization of installation offices should not lie still and allowed for continued work between GEMA and Elac but under Elac's control.

With this decision they discredited the long remembered contributions of GEMA in the rapid realization of Kühnhold's ideas and knowledge in the fields of both sound and radio location. They especially derided the work of Brandt and his co-workers, who had quickly acquired for themselves and GEMA notable positions in this field through their audacious designs.

After the successful introduction of the new standards of sonar, Brandt had carried out further improvements. On the basis of his work GEMA could offer the Navy a whole series of developments that defined them as the master enterprise in underwater sound. Among other things they developed (1) the GH console with eight cylinder-mounted sweep mirrors for high accuracy in range and for short-range targets of about 300 m; (2) the hearing aid that allowed the operator on the basis of the Doppler effect to distinguish between stationary and moving targets and to be able to estimate relative speeds; (3) the Doppler tachometer for measuring the relative speed of a target to within a knot; (4) a very small sound pulse generator of 3–5 kW power; (5) the compensator for stationary sound-locating devices with fixed microphone mount and for mounting in a bow-bulge; (6) the duplexing circuit using neon bulbs or diodes for disconnecting the receiver without a relay during transmission; (7) the console with CRT and electrical time base and (8) the twin amplifier for the console using Wehrmacht tubes as a heterodyne twin and with gain as a function of range.

GEMA was not content to allow itself to accept a position in sonar techniques inferior to Elac.

At the same time GEMA received the communication just mentioned from Naval Ordnance they were also openly reproached by Naval Command that, because of their increased acceptance of Luftwaffe orders, they forced the Navy to give more orders directly to AEG. That was a definite sign of a shrinking fidelity in this sector as well. Now only GEMA's management was in the position to take measures to counteract and alleviate the obvious worsening of their business relations with the Navy.

After internal deliberation among their closest circle von Willisen and Erbslöh came to the conclusion that it would be useful and wise to accept Navy Fiscal Office participation in their management. This decision, made grudgingly only for tactical reasons, became easier because the reasons impelling the Navy participation in management lay entirely outside their business interests. Naval Command had no economic interests, only a war-dependent interest in GEMA relative to their production. A delegation headed by Erbslöh was able to assert GEMA's convictions over those of the Fiscal Office.

The naval interests should have been represented as a fully authorized business partner that held a third of the shares. GEMA had the option to recover these shares within a year after the end of the war, preserving the company for Erbslöh and von Willisen. The Navy stipulated as necessary for their interests that they reserve a decisive control over the majority in determining GEMA's development and production. The profits were, however, to accrue to the majority share holders. The Naval Fiscal Office was to receive a guaranteed dividend but no share of the profits.

In a business meeting on 17 February 1943 it was decided to increase the common stock from 1000 000 RM to 1500 000 RM ($235 000 to $350 000), with the Naval Fiscal Office contributing the difference. This cleared the books of the loan made earlier by the Naval Command. On 4 March a board of directors was formed that replaced the board formed in 1936. The new board consisted of Lieutenant Commander (retired) von Simson, Chairman; Planning Officer (Naval Ordnance) Schüler, Vice Chairman; Defence Economic Leader Erbslöh; Lieutenant (reserves) von Willisen; Admiral Gladisch, Naval Technical Department; Attorney Dr Veith.

Erbslöh and von Willisen were at the same time GEMA managers together with deputies Golde and Lassen.

In July 1941 Naval Command involved themselves in a patent dispute between GEMA and the Atlas Company. They prohibited GEMA from pressing its claim for the duration of the war and ordered them to conclude a licensing agreement with Atlas. GEMA's complaint turned on a revocation of one of Atlas's patents from 1927. GEMA had employed some of its essential characteristics in their Messkette. The objection hurt GEMA financially, and they did not follow the wishes of Naval Command and complained, despite wartime conditions.

The patent concerned problems of travel-time compensation in devices for determining the direction of noise. The point was that it covered the equalization of time and phase differences through circuits made up of elements having identical time constants. In order to equalize delays in a binaural listening technique for sound location, chains of substantial length were required. According to the disputed patent a reduction in the number of elements in the chain was made by constructing it from coarse and fine steps. The patent covered the concept of compensation for time and phase differences by the coarse–fine division of the

126

electrical delays. GEMA questioned the amount claimed by Atlas for inventor's rights and triumphed in a decision of 25 March 1943.

In mid-1941 GEMA recognized the inadequate capacity of their plant on Wendenschloss Strasse in Köpenick to handle the requirements for radar and sonar equipment, and by the third war year there were limits to the availability of subcontractors, as it was difficult to find free industrial capacity on the market. No one was ready to undertake orders from GEMA, all were fully occupied with war orders. Owing to the confined space an appreciable increase in the current production capacity was not possible. The fabrication of rotatable bases took a lot of space. Part of the warehousing for completed equipment had been moved to rented rooms in Berlin to make room for production, and this caused the loss of time and money.

In August 1941 GEMA rented 110 000 square metres of land from the city for the duration of the war on Achenbach Strasse (today Salvador-Allende Strasse) not far from Wendenschloss Strasse. There they built rapidly with light construction a second plant, which they began to occupy in early 1942. An increase in personnel was only possible primarily from women and unskilled labour conscripted for essential service. Inasmuch as only limited numbers of competent workers could be drawn from production to train new people it was necessary for laboratories and experimental shops to contribute to this. The continual drain of personnel from these areas hurt because there were already deficiencies in the production of test and measuring instruments that resulted from them having to construct special equipment, the completion of which required in 1941 about 800 000 man-hours.

The first series production of the new Freya-LZ, FuMG(Flum)42G(fZ), was celebrated in November 1942. This radar, whose base had been developed in co-operation with Luftschiffbau Zeppelin, represented a significant improvement for deployment by the Luftwaffe in locations where frequent change of station was not required. The base was constructed from standardized units. Four frame arms served as the supporting elements arranged at 90° from one another from the centre. At the end of each arm was a levelling screw with a plate that allowed the rotatable base to be given solid support. If necessary, the ends of the frame arms could be screwed to a prepared foundation. The massive bearing with attached gear-ring was placed at the centre on the platform with the bearing and drive mechanism. A 12 m long mast was assembled for holding the antennas and was screwed to the bearing housing. Around it the cabin was built on the platform. The first version of this cabin, whose construction GEMA had initiated already in 1939, was circular (but later square). In series production the cabin was made octagonal to reduce wind resistance.

In the cabin of Freya-LZ the electronic elements R, P, O, Z and N were arranged conveniently to the left and right of the mast. At the rear, now protected in the cabin, was the transmitter unit T and IFF unit Q. Unit P for lobe switching was now incorporated as part of the basic set. Cross-pieces of the frame construction served to hold the dipole arrays. The antennas were raised to the proper height by cables running over a bracket at the top. Horizontal slewing of the base was done by a velocity controlled transmission or by hand. The direction was read on dials. The 380 V, three-phase current was either drawn from the power distribution net or from an engine-driven generator. The range of the Freya-LZ, later given the designation FuMG 401, with its new 35 kW transmitter was 200 km. When GEMA introduced a transmitter with 100 kW, it was necessary to extend the Messkette from 200 to 400 km with the auxiliary OZ(OKZ).

The Luftwaffe had excellent results from the first experimental deployment of a Freya-LZ as a vectoring radar for two Würzburg Giants, which had to locate the hostile aircraft and the attacking fighter for ground-controlled interception. The Freya picked up the incoming target early and transmitted its bearing to the Würzburg for determining its altitude. Because the range of the Würzburg was shorter than the Freya, it had to look in the direction of the approaching aircraft in order to pick it up as soon as it came within range. With the ever-increasing numbers of bombers, the night fighters felt pressure for a greater rate of success, which led to increased demands for Freya-LZ. The construction of this radar, of which barely 400 were produced by the end of the war, remained with GEMA, although for a time Lorenz made bases. The Freya-LZ was referred to by the British as the 'Pole Freya'. The earlier version with hydraulically lifted antennas was called the 'Limber Freya', the result of it being towed like an artillery limber.

In 1942 GEMA could not expand their production, even with the help of AEG and Telefunken, to the degree that their customers wanted. They were able to increase returns 35% over the previous year to 47 000 000 RM ($11 000 000). During the year they delivered 333 sonars with the new indicators but 99 had to be delivered without the A/B module, which AEG made. They delivered 502 radars, 184 f-wave and 318 g-wave. The number of employees grew to 3657.

CHAPTER 18

ON THE WAY TO POWERFUL RADAR AND SONAR

As GEMA's workforce returned to their benches on the first day of 1943 many suspected that a year of privation stood before them. The war made itself noticeable to them not only from the separation or death of family and friends but also through the steadily increasing air attacks that made the war a local as well as a distant reality. Berlin shared with other cities the fate of being the recipient of more and more bombs, and the arrival of enemy aircraft brought the demand for more effective defence.

It was obvious that GEMA's plant, as an indispensable supplier for air defence, had to take special precautions for its own protection. During summer 1941 anti-aircraft guns were placed on factory buildings and on an adjoining playing field. The Flak was part of a battalion in Neukölln; the platoon leader, his headquarters and command were housed in barracks on the playing field.

The expansion of the war toward the end of 1942, the time of the great offensive against Stalingrad, caused the Flak platoon to be sent to another assignment, and GEMA had, against their will, to be content with a home-guard battery composed of soldiers unfit for service at the front and of auxiliary personnel. Because of GEMA's secret character and activities many of these personnel had to be recruited from the plant. As the war between radar sets raged almost as intensely as the war between guns, demands for secrecy were driven to unparalleled extremes. Over 170 workers born between 1901 and 1922, who had been exempted from conscription owing to their high technical qualifications, were ordered to serve in the Flak batteries protecting the factory and were sworn in as soldiers. Active service in air defence took priority over their industrial duties. Action could take place at any time, but there had to be at least six hours rest between Flak service and radar production. Many were thereby held back from their important work by the formation of four work shifts to accommodate the situation.

The introduction of the home-guard Flak and the enhanced wartime regulations for maintaining secrecy led Erbslöh, as 'captain of the good ship GEMA', to evaluate and examine the significance of these measures for which he had little appreciation. It also went against his humane convictions that the increasing number of prisoners employed were to be treated differently than their German co-workers. He held that a good working climate and an enjoyment of the job were requirements that applied to anyone working for the company, regardless of who he was or where he stood in the organization. Von Willisen, through his business and personal connections with Naval Command and to the Air Ministry, not to mention his position on active duty, secured some limits on the security forces overseeing the company. Both hoped that they would remain unmolested, especially now that the Fiscal Office was participating in the management.

Along with the necessary expansion of the night-fighter control came the obvious and pressing need for the Flak's radar to be able to distinguish between the fighters and the enemy bombers. The Luftwaffe had hurried too much in instigating procurement of Telefunken's IFF for the Würzburg early in 1940. This device differed from GEMA's IFF Stichling in using the radar pulse to interrogate by altering its pulse repetition rate from the normal 3.75 kHz to 5 kHz. This interrogation procedure was selected in order that the IFF would reply only through the aimed command of the radar operator and would not return signals continuously. The interrogated aircraft returned a signal of 1 W power at a frequency in the vicinity of the Freya f-band. The IFF transmitter was modulated with an adjustable audio-frequency Morse-code signal that was heard by the radar operator on a receiver, 'Steinzeige'. Because of its identical shape this IFF set received the name 'Zwilling' (twin). It proved to be quite impractical and had to be modified repeatedly by the troops, but these alterations were not sufficiently successful to allow it to be used; meanwhile 20 000 units were left to sit unused in a warehouse.

Now GEMA's flexibility showed itself. At a time when they were heavily over-committed they were able in short order to redesign their own IFF in such a way that essential parts of the idle Zwilling could be used. In order not to have to remove the mechanism that was already in the planes, they altered the GEMA IFF so that it fitted in the same frame, and at the end of 1941 they presented their reconstructed IFF to the Luftwaffe. It had already been tested in its original electronic form and was quickly accepted as FuG25A with the cover name 'Erstling' (first born).

In mid-1942 GEMA received the order for a large series of FuG25A, and by the start of 1943 the Luftwaffe expected monthly shipments of 3500 sets. At the end of 1942 GEMA began with 700 monthly, quickly increased to 1500. In order to meet these demands subcontracts had to be awarded to Blaupunkt, Roland Brand and Tekade, which allowed the required number to be attained in March 1943. Of course, at the same time production also had to begin with the ground receiver Gemse. Thus all radars that were used by the Luftwaffe were equipped with the IFF FuG25A. The Telefunken radars, which worked on 53 cm, had to be additionally equipped with the interrogation transmitter Kuh. GEMA produced the ground receiver Gemse but awarded the fabrication of the interrogation transmitter to subcontractors. By mid-1943 these IFFs were in general use by the Luftwaffe. A complication arose when GEMA received 20 000 Zwilling sets from the Air Signals Arsenal in Pankow in early 1943. Cannibalizing these sets and separating and arranging the components for new equipment added to their burdens. In May the Luftwaffe made it known that they would be increasing their total IFF requirements substantially, and GEMA transferred this order for 25 000 sets to Siemens. In accord with their agreement with GEMA, Siemens obtained a direct contract.

In order to produce the basic electronics of the new air-warning set Freya-LZ (Pole Type) under their own management, GEMA shifted production of the earlier set (Limber Type) to AEG and Telefunken. This was necessary because, despite the expansion of the shops on Achenbach Strasse, there was simply not enough room to assemble them. And the equipping of vehicles necessary for maintenance and repair required space.

With every additional subcontractor, GEMA unavoidably required more employees just to supervise them, but the efficient co-ordination of external fabrication brought a significant increase in production in the first half of 1943. It is worth mentioning that at the time the subcontracting firms were experiencing damage from air attacks, although GEMA itself had as yet not been seriously affected.

Even in the fifth year of the war the further development of new techniques and equipment was not cut too severely. Brandt and his people in the low-frequency laboratory had avoided no pains in revising the design of the Messkette to improve its accuracy and reliability while reducing its fabrication costs. Testing new Messketten was carried out by means of a very accurate calibration device. In addition there was a great success with test bench K, Type PTK-2, which Brandt had developed for use in assembly shops and field stations of the Navy and Luftwaffe. By mid-1943 80 of these test benches had been delivered, along with many other specialist test instruments, in order to ensure that there would be a sufficient number of accurate instruments available. In addition to all of the normal instruments and signal generators needed for the Messkette test bench PTK-2 there was a standard Messkette for 0–200 km corresponding to phase shifts in the range of 0–240° at 500 Hz pulse repetition frequency. Its measuring range was divided into coarse steps of 10 km and fine steps of 50 m, accurate to better than 0.1%. A Messkette to be checked or calibrated was connected to the test equipment where it could be checked against the standard Messkette. The test bench also served to check the audio oscillator ZP and the two CRTs, NB and OB. GEMA made an equally expensive test set PTN-1 for the exact testing and aligning of radar receiver N, which they delivered in equal numbers to PTK-2. GEMA provided these all-encompassing measurement instruments not only to their own assembly shops but also to the workshops of the users to give them the capability of keeping their radar in top condition.

In practice the Messkette with step switches, which GEMA had constructed for reliability in long, continuous operation, was showing itself to be less than ideal, the consequence of contacts that were not wear-resistant and of small jumps made in the transition from coarse to fine. **In early 1943 Brandt started the development of a circuit that had contacts uninterrupted by coarse–fine transition switches. The solution consisted of a circuit in which 10 km links of the delay line were sequentially probed with a differential rotary capacitor. In its practical form two such capacitors were coupled to the drive mechanism and the range scales and successively connected to the input of the deflection amplifier. With an ingenious relay circuit the stators were connected one after the other to diverse 10 km network links. In the same way the rotors were connected one after the other to the deflection amplifier input. With these each 10 km range could be continuously probed and the jumps of the transitions were avoided. This solution showed itself to be a fundamental improvement of the Messkette not only in reliability but also in ease of operation. Trials of this Messkette led to its final design and production.**

A disadvantage of the new Messkette was that the alternating connection of the two rotary capacitors had to be made with two network links, although these no longer needed fine steps. In order to avoid the need for using two lines, Brandt had a differential rotary capacitor with three stators built. It allowed a circuit to be constructed in which the outer stators were sequentially switched to the elements of a single network. This sequential Messkette was the best that GEMA was able to offer for range measurement. Its accuracy over long periods was limited only by the stability of its components. Circumstances prohibited further development of this approach to the problem of the Messkette.

As a consequence of the increasing air attacks on German cities, and especially because of the fear that GEMA could be hit, Naval Command ordered on 7 April 1943 that the entire radar research and development department and a part of the fabrication facility must be moved to a region thought to be in less danger within the shortest possible time. This move had the agreement of Speer and the Air Ministry. From the two locations that GEMA was offered, they elected to relocate the research laboratory, the assembly area and the entire prototype construction to a former cloister in Wahlstadt near Liegnitz, which had been used as a cadet school and, during the war, as a prison for captured enemy officers until early 1943. The grounds had an area of 72 000 square metres with buildings on 11 000 square metres. The move from Köpenick to these new quarters, which carried the cover name 'Blücherwerk', began on 1 July 1943.

GEMA decided to use rooms of a branch of the firm of Gustav Winkler in Lauban, Niederschlesien for distributing part of their production. This location had grounds of 23 000 square metres with buildings on 11 600 square metres. On additional ground rented from the town of Lauban GEMA constructed a large assembly hall of 2400 square metres as well as six houses for quarters, a community centre, a clinic, a small administration building and barracks. The cover name for this location was 'Rübezahl-1'. At the end of 1943

GEMA rented from Winkler yet another plant of about 5500 square metres, 'Rübezahl-2'. Transfer of manufacture to Lauban took place gradually because the space was given up hesitantly and after it had been converted to its new uses.

These measures, which Erbslöh and von Willisen had expected, overloaded GEMA significantly. The costs of the move, which were only partially covered by the Reich Fiscal Office, hurt, but the loss of time was worse. Especially hard hit were the branches of development and construction, which had to be moved completely, and Erbslöh had to divide his management time in bringing all the different parts back into a coherent whole. Brandt was named in recognition of his services and value as a deputy of the firm and made leader of the Blücher plant in Erbslöh's absence. Brandt remained the leader of the low-frequency laboratory and continued to influence research in his characteristic manner.

In the field of underwater sound the Navy worked out their needs with Elac in Kiel, which did not hinder GEMA in carrying out their independent research. The circumstances of the war forced the two companies closer together anyway in order to overcome common problems, and it did not trouble GEMA that they had to harmonize with Elac.

Because the development of more compact sonars with CRT indicators—the so-called SR sets 'Nibelung' and 'Notung'—was delayed, an interim solution was followed in re-working the older S sonar sets. These interim types, called SZ, had essential improvements. The twin amplifier in the console was equipped with gain control determined by the travel time. A heterodyne unit for detecting the Doppler effect with headphones and a graphic recorder were added. Relays were replaced by electronic switching between transmission and reception with the new quadrupole transducers of Atlas or Elac.

In order to make the S sonars interesting again for U-boats GEMA developed a small SU set with the cover name 'Mime'. **In the upper compartment of one rack were the newly developed small twin amplifiers that used standard Wehrmacht tubes and had delay-time gain control. It also had an integrated audio unit to detect Doppler effects with headphones and had the power supply next to the indicator (Trommelkino), newly developed by Preiss. This novel indicator was equipped with eight sweep mirrors on an octagonal turntable with the CRT in the centre of it. The speed of the turntable was governed and allowed four sweep ranges: 300/900/3000/9000 m. In the lower compartment was a compact, newly designed sound generator with capacitor for 5 kW pulse power and its power supply.** The volume and weight of Mime was only half as great as that of the normal S sets. In mid-1943 GEMA delivered eight prototypes for testing by the Navy, but since in the meantime the priority for radar for U-boats became greater than for sonar the tests were postponed. Production was originally scheduled for Lauban, but GEMA never received a contract for delivery.

In the radar sector came the demands for new developments. GEMA gave the Wismar modification of existing equipment for changing frequency to subcontractors and depended on Siemens for the delivery of tunable Freya-LZ transmitters, who provided trombone-like circuits that could be tuned. The Freya receivers required only a minimal alteration, as the original design was sufficiently good that a modification using newly developed tubes was not worth the effort. Schultes had begun working on new receivers that could replace existing ones simply by exchange since his transfer to Siemens.

The urgent need for extending the range of all radars put pressure on the high-frequency laboratory to increase transmitter power. Neither for mobile ground radars nor for ship-borne equipment was it possible to increase the area of antennas as had been done for Mammut and Wassermann, which had attained greater ranges with the usual powers. Increasing the power of the existing transmitter designs had been accomplished with the powerful output triodes TS60 for g wave and TS100 for f wave, and development along those lines was essentially complete, but high-power pulse modulation for a megawatt or more presented problems.

By the end of 1943 there were prototypes ready of the 400 kW modulator 'Eber', in which a capacitor was discharged through a modulation transformer by a thyratron. In it the two TS41 output triodes of the Freya transmitter were modulated with 3 μs pulses at 25 kV. The power achieved was 100 kW. The modification of existing sets with Eber could be made by replacement of certain components, but a general modification of existing sets, however, did not take place. It was thought better to collect experience first with this modulator because the use of high anode potential had uncertainties. A few Mammut-Friedrich and Wassermann sets were nevertheless quickly changed to this mode of operation.

The call of the Navy for increasing the range of their Seetakt sets, for both surveillance and gun-laying, was more emphatic than that of the Luftwaffe. Their g-wave sets, which used grid-modulated transmitters with two decimetre triodes of type TS6, gave a maximum range of 30 km and obviously required increased power. The development of the essential output triode TS60 had been completed at the start of 1943, so production could begin. Siemens took over production of the new TS60 because GEMA's tube production for the new radars and for the supply of replacement tubes was over-committed. By mid-1943 the development of the transmitter module 'Gisela' with high-power tubes was completed. This allowed power on the g band to be extended to 125 or 150 kW. In combination with a partial increase in antenna area where this was possible, this led to ranges with ship-borne sets of up to 60 km, according to the size of the target ship. Even from U-boats the increased transmitter power gave ranges of 20 km against sea targets, despite the low position of the antenna. U-boats were in the meantime being attacked increasingly from the air, so at the beginning of 1944 GEMA designed a 100 kW f-wave anode-modulated transmitter for the air-warning set 'Lessing', which radiated omni-directionally. A vertical dipole mounted on top of the conning tower was used in common mode for transmission and reception. The range accuracy for aircraft or ships was only 5%, which was good enough for a warning. The range against aircraft was 25–30 km.

The department for special fabrication, to which the production of prototypes belonged, was hit hardest at the time by the move from Köpenick. The series of special test components with the designation EA had been extended to EA23 'Hornfisch'. These sets carried codenames generally selected randomly from the names of fish, e.g. 'Barsch' or 'Lachs'. With the increasing numbers of radars and sonars the number of instruments and test equipment needed for their maintenance and repair also rose. Because of the peculiarities of their fabrication and the small number of them produced, the needs for special devices were poorly combined with those of series production. The production of the IFF set FuG25A, Erstling, and its accelerated installation in aircraft that had been prepared to take the abandoned Zwilling made the need for the support instrument FuP25A, 'Erstwart', desperate.

Since the group in Berlin engaged in the fabrication of test instruments, as well as a whole series of other devices that were not amenable to series production, had to enlarged, it was among the first transferred in August 1943 to the Blücher plant in Wahlstadt. Because the activities of this group could not be interrupted, they ended up larger at their new location. For this growth only a few of the crew in Berlin could be made free, and the main burden fell, as it had before, on the research and development group and their associated shops. They transferred quickly in autumn 1943, but an interruption of the work in progress was unavoidable during the time of the move. It was astonishing that already in October, even in the unfinished working space, the urgent development of high-power anode modulators could be continued.

In order to increase the power of the thyratron modulator to a value substantially higher than 500 kW, GEMA's tube laboratory produced a series of gas-discharge tubes with cold mercury cathodes and spark-needle ignition for high instantaneous currents. The work was based on the experience gained in the development of similar tubes for underwater sound. For a laboratory model of a big 80 cm transmitter with two TS60 triodes, a modulator was constructed with 2 μs pulse length and a duty cycle of 1:10 000 that delivered more than 1 MW from which the transmitter delivered an 80 cm pulse of 400 kW. In order to have the smallest scatter in pulse initiation, GEMA used 12 parallel-switched needles in the discharge tubes. The trials showed them to have a lifetime of only 60–80 hours with an unavoidably large scatter in the ignition time, which would require special measures to counteract.

In great haste GEMA made a special stabilizing circuit, 'Specht', for synchronizing the time base in the indicator and the transmitter pulse, whose initiation was scattered about a statistical average. This circuit was needed because the transmitter pulse, which had to initiate the CRT traces, could not be accurately triggered by a central-synchronized fixed-phase system. The Specht circuit allowed ranging with the time base calibrated in distance units corresponding to deflection time. Fine ranging with the Messkette was no longer possible.

But the demand for better resolution with the Messkette could not be met with such correction circuits, and GEMA had to retreat from their high standards of accuracy. In all sets where the high-power modulators were not necessary, the Messkette remained sufficiently precise. Ranges determined with anode modulation accomplished with low-scatter thyratrons (those ignited by grid control rather than by spark needles) were satisfactory for exact gun-laying. For the larger ranges that could only be obtained with high transmitter power (sets for which it was not possible to increase the antenna size) it sufficed to read the range from the CRT without recourse to the Messkette. GEMA fulfilled this requirement with the range measuring device 'Paris'. The transmitter pulse triggered a linear time base, which was independent of the fluctuations that caused the time scatter: it used pulse repetition rates of 50, 100 or 500 Hz.

With this time base in 1943 GEMA developed the range circuit 'Paris 0' and 'Paris 1' for various distances. In 1944 they added 'Paris 2' that was based on the invention of a delayed-pulse generator and that could display selected portions of the range trace on the screen. This circuit was combined with the new transmitter 'Boulogne' that used a spark discharge for anode modulation to attain 400–1000 kW. The radar from this combination was called the 'Paris–Boulogne' and was used for sea search and air warning.

At the end of 1942 GEMA entered a new field of physics when Preikschat's group initiated research in using infrared for determining direction. The project received the cover name 'Froschauge' (frog's eye). Its object was a passive direction-finding set that allowed an object to be detected and examined. **The transformation of an infrared image into a visible one through special optics was made with a suitably small photocell of lead sulfide mounted on a thin, spring-mounted piece of glass. Two electromagnets, oriented at 90° relative to one another, moved the cell. The alternating current in the two magnets differed slightly in frequency relative to one another so that the small cell followed the path of a changing Lissajous figure in moving to the collector. The amplified signal from the cell modulated the beam intensity of a CRT, thereby forming a visible picture from the pattern traced by the CRT beam in following the motion of the cell.** A sharp image was formed from a combination of a lens and a focusing mirror. The image converter was given the cover names 'Grobbiber' or 'Biber'.

The path to correcting the problems that were attached to Froschauge was tedious and proved to be long because the project lost priority not only to radar development but also to work in countermeasures being undertaken to determine what kinds of radar the enemy was using and how to interfere with it, as the radar war had broken out in full force. The transfer of the laboratory and test shops also took longer than expected. A laboratory version of Froschauge finally came to be tested in autumn 1944. It showed that determining the direction of distant targets at sea was not possible. It produced a good picture in wide angle for close-in objects, but for distant objects the weak signal voltage from the cell was overcome by microphonics. Various attempts to improve the vibratory motion imparted to the cell failed, and after a while GEMA gave up for good any further attempts, although there were ideas that had good prospects.

GEMA had a great triumph in 1943 with Freya-LZ, as it found unanimous praise with the Luftwaffe. All their previous experience found its consummation in this radar, so that troops did not have to fight with the irritating deficiencies that marked inadequately tested equipment. The early mobile sets (Limber type) had been in service for more than three years by then, so that the early defects had been corrected. Sets with lobe switching and IFF, features incorporated into the basic design, had been well tested and problems eliminated. Intensive discussions and thought had led to the conclusion that reliability was not to be sacrificed for the gains of using a common antenna.

Just as with the mobile rotatable bases the lower array was used for transmission, the upper for reception, both mounted above the electronics cabin. The IFF antenna did use common-mode operation, even though the transmitter and receiver worked on slightly different frequencies. The make-up of a Freya-LZ varied only slightly from one set to another.

The standard model with an operating cabin had Z (audio oscillator and phase-shifter for T and Q), R (transmitter control with modules RH, RN, RI), T (radar transmitter with modules TU, TS, TN), FM (frequency meter), V (radar transmitter antenna), W (radar receiver antenna), LM (power meter for T), PU (receiver lobe switcher), N (radar receiver with modules NE, NB), O (range indicator with modules OK, OB) and OKZ (supplement to Messkette).

The IFF make-up had Q (IFF transmitter with modules QU, QN), VW (IFF transmitter–receiver antenna), FM (frequency meter for Q), LM (power meter for Q), PU (IFF lobe switcher) and P (IFF lobe-switching indicator).

GEMA turned over production of the older model Freya entirely to AEG and Telefunken, while they undertook the production of radars of the new LZ construction with Lorenz. The Luftwaffe increased the quota for Freyas substantially in 1943.

To the Navy GEMA delivered the ground radar Calais for coastal watching and 142 sets of FuMG-Flak, which they had developed in parallel with the Würzburg for naval Flak. The radars for U-boats were sluggishly produced, but by the end of 1943 GEMA received welcome competition from Lorenz that relieved them from this pressure. Lorenz had built their 'Hohentwiel', an airborne sea-search radar that was small and light enough to make a good set for U-boats. For ship-borne radars a collective name 'Zerstörergeräte' (destroyer set) was applied, because the antenna mounting was on the 'Zerstörersäule' (destroyer column) made by the firm Beeck in Oldenburg, and which was mounted at some appropriate place on the ship. The electronics were lodged in the ship's radio room. The Navy provided a variety of slightly different antennas for these radars.

The eventful year 1943 unexpectedly ended successfully. Despite the disagreeable conditions brought by the war, and especially by the difficulties imposed by the move, production increased significantly over the previous year.

First-generation sonar production without generators increased to 100, 101% over the preceding year, and 322 with generators, a 140% increase, but it was in radar where production really grew. Production of Freyas went to 584 sets, a 318% improvement. The Navy received a total of 487 assorted sets for a 153% increase over the previous year.

Aside from IFF sets, which suffered large interruptions in deliveries owing to the bombing damage suffered by subcontractors, GEMA fulfilled the quota set for them in 1943 by Naval Command and the Air Ministry and by the Radar Working Committee, although in various meetings credit was not given to them. They were even able to fulfil an enlarged quota on instruments and test equipment, which had required in the meantime an increase in capacity. Because of modifications and improvements they were 13 vehicles behind the 58 ordered for maintenance and repair to the Luftwaffe; similarly only 17 Wassermann sets instead of the 37 ordered were delivered.

For the year 1944 GEMA planned substantial improvements in various items in their line of products. In first place was the increase of radar transmitter power. To provide support for their research staff a greater co-operation was planned with scientific departments at universities and technical institutes where competent help was to be found. Preliminary discussions had been held and the outlines for a contractual collaboration already worked out in part. Contracts were also planned with Naval Ordnance to support improvements for anode modulators 'Keiler' and 'Büffel', for the large transmitter 'Grete', for g wave 'Wismar' and for prototypes of 'Boulogne g and c' and 'Dünkirchen g and c'.

At the beginning of 1944 questions concerning the relationship that Kühnhold had to GEMA during the early years were resolved. Some of GEMA's important patents and concepts rested on his ideas and proposals, and much early work had been done with him. Because no special descriptions had been kept of the origins of GEMA patents, it was difficult to determine in 1943 what rights belonged to him. In order to compensate him for his part in GEMA's profits, an agreement was made, with the consent of Naval Command, for a financial consideration. They obtained from Kühnhold, for a fixed sum which had the effect of a purchase, his rights as inventor or co-inventor. For the payment of an additional sum they came to an understanding about license rights that took effect on 1 January 1944. These obligations were to remain in force until ten years after the end of the war and were independent of any time limits on patents granted to Kühnhold. With this GEMA received unlimited and exclusive rights for all patents, registered or pending, which could be traced to contributions by Kühnhold. Except for licensing fees, which were not paid on sales to the Wehrmacht, they had by then covered all claims that might come from Kühnhold.

The successes that GEMA had achieved for 1943, and the optimism despite the war with which they entered 1944, were overshadowed by the bombing the Köpenick plant received from 3:30 to 5:00 AM on 24 December 1943. On Wendenschloss Strasse all buildings had extensive air-pressure damage to roofs, windows, doors and walls. Some buildings had to be evacuated and could be reoccupied only after time-consuming repairs. Some buildings of light construction were destroyed, and others burned completely or partially. Damage to the second plant on Achenbach Strasse was limited, arising primarily through fire that resulted from a large amount of packing material that had been stored in a large tent on the grounds. Erbslöh's house on the grounds was made uninhabitable and the two dwelling houses on Luisen Strasse and Marienlust were destroyed by fire. Now the advantage of dispersing important parts of the plant to places not in danger of air attack became apparent.

CHAPTER 19

INTENSIFICATION TO THE LIMIT

A first examination of the bomb damage from December showed that GEMA had had luck in its misfortune. The buildings used for testing, assembly and research had received the most serious damage. As a result of moving almost the entire laboratory to Wahlstatt, the development activities remained intact, but production of rotatable bases for the Luftwaffe and Navy was reduced by 20% as a result of the damage received by the assembly hall. Within a few weeks this had been made up by an acceleration in the use of the plant in Lauban, with the prospect of an increase in production.

Orders reached a level at the beginning of 1944 that was markedly higher than for 1943, despite the increasing obstacles brought by ever more serious war conditions. There were naval orders for 753 Seetakt sets of Type 501 for U-boats and over 254 of Type 107 for surface vessels. In addition there were 262 orders for delivery to begin in mid-1944 of Type 150, the Flak set with lobe switching for elevation. For the Luftwaffe there were orders for almost 600 Freyas in assorted forms but primarily for Freya-LZ.

A number of improvements were foreseen for various elements of naval radar in 1944 with an increase of transmitter power the chief goal. Already in 1943 a whole series of alterations and extensions were undertaken on all radars. Changes were needed to protect their operation from enemy jamming by giving the equipment a variety of wavelengths, which compelled substantial alterations. Inasmuch as no way was immediately found to provide existing equipment with a wide range of tuning, a provisional solution was taken by distributing modification kits of so-called 'islands' of small frequency bands. Because of bottlenecks in their own laboratory, GEMA had transferred the development of tunable f-wave transmitters to Schultes at Siemens, which complicated matters.

It was clear to GEMA that the increase in production demanded for 1944 would be hard to meet. The transfer of part of their construction to Lauban must, even under the most careful planning, bring with it a costly loss of time. They had to move more than 50% of all machines and shop facilities and almost the entire aggregate of their toolmaking, coil and transformer production and the galvanizing shop. Through a protest to Speer's ministry Erbslöh was able to prevent an even greater proportion being transferred to Lauban, as he feared a retreat on the Russian Front, even in the latter half of 1943, which could compel a return to Berlin. Furthermore, the deficient and delayed deliveries in the allocation of supplies, the insufficient and at times hopelessly delayed deliveries by subcontractors as a result of air attacks, the general deficiencies in the control of industrial effort and the personnel shortages made things especially difficult for him.

By the end of 1943 GEMA had 5460 employees, of which 4003 were still at the main plant in Berlin. The staff in Wahlstatt numbered 358, in Lauban 1099. About half of the personnel at Lauban had been taken from the Berlin works. Personnel assigned to the fabrication of equipment or the constituents thereof that could not be given to subcontractors, either because of the nature of the work or for reasons of secrecy, amounted to 1560. Subcontractors employed more than double this number in producing GEMA equipment.

The ever-growing need for radar and relevant supplies brought ever-increasing governmental orders to GEMA, orders that exceeded their capacity, forcing them increasingly to have work done by very small firms as well as the large ones used until then. As the head of a widely scattered production complex the company had accomplished the following tasks: (1) fabricating prototypes to maturity, (2) controlling the subcontractor's production, (3) correcting flaws in the subcontractor's production, (4) modifying the subcontractor's products because of changes in design, generally caused by the purchaser, (5) assembling the subcontracted work into a complete set, fabricating cables and auxiliary parts and (6) testing the completed set mechanically and electrically. These tasks required a tight organization, which became harder to maintain with every day of war.

The relocation of the entire development and experimental construction to Wahlstatt had been beneficial. In Berlin this department had been required to furnish replacements for plant technical people and ever more often to aid production workers in their tasks. Many development projects had been delayed through this. The pleasant environment in Wahlstatt, compared with the suburbs of the capital, which at that time were the recipient of frequent air warnings, allowed for continuous work on their many research projects.

From the beginning of their engineering careers Erbslöh and von Willisen had drawn eagerly on the advice and help of various scientific institutes, especially on the Heinrich Hertz Institute for High Frequency Research during the early years. This institute had provided them with two competent scientists, the physicists Dr Schultes and Dr Brandt, who created GEMA's high- and low-frequency laboratories and who attracted a number of other scientists from all corners of Germany. By mid-1943 it was clear, however, that GEMA's scientific capacity no longer sufficed to solve fundamental problems in radio location and associated areas, which presented the clear need for a stronger collaboration with the relevant institutes. Owing to the war conditions these problems might have best been solved in alliances with other companies that were active in radar, but GEMA chose to preserve its independence by granting research contracts to institutes directly.

Earlier efforts to bring in outside scientific consultation, especially on basic matters, had always failed due to prohibition by the Navy, who cited security regulations. Kühnhold had supported GEMA with his scientists when that was required, but in the meantime the tasks confronting both parties had grown to the extent that there was no time for collective effort. With time, however, both the Naval Command and the Air Ministry had allowed GEMA to place research contracts with physics institutes in which the needs of secrecy were not normally habitual. This allowed them to carry out GEMA's research wishes relatively unhampered.

In 1944 for various projects, but primarily in the areas of high-voltage-pulse techniques, broad-band antennas and long-persistence CRT phosphors, GEMA granted research contracts to (1) Institut für Starkstrom- und Hochspannungstechnik (heavy-current and high voltage) der TH (Technischen Hochschule, technical institute) Dresden, Professor Binder, (2) Hochspannungsinstitut der TH Hannover, Professor Schering, (3) TH Darmstadt, Dr Fischer, Dozent (assistant professor), (4) TH Karlsruhe, Professor Thoma, (5) TH Munich, Dr Thomascheck, (6) Physikalisches Institut der Universität Wien, Professor Stettner, and (7) Institut für Radiumforschung in Wien, Professor Ortner.

In early 1944 the firm Geräteentwicklung Danzig GmbH had been brought in to develop high-power modulators. Among other things this company, which had specialized in pulse electronics, received a grant to develop a modulator of up to 100 kV for the newly developed triode TS100 for a high-power f-wave transmitter. In collaboration with Professor Erwin Marx of the High-voltage Institute of TH Braunschweig they developed spark methods for discharging a capacitor or a high-voltage delay line, which operated with a 50 Hz pulse frequency and yielded a 1 μs pulse width. Peak powers of 2–4 MW had to be generated. This modulator gained the internal name of 'Funkenspule' (spark coil). Two other modulators of similar power were built for 100 Hz operation and carried the names 'Trafobüffel' (buffalo transformer) and 'Trafokeiler' (wild boar transformer), Development work was undertaken to build modulators for 500 Hz, 'Bulle' (bull), and 650 Hz, 'Saurier' (crocodile, dinosaur).

The difficulties of high-modulation voltages in combination with anode modulation were new and unusual for the GEMA engineers, and they were thankful for the help they received from other firms and institutes that had had experience in the application of the highest pulse voltages. In operation, radar sets experience extremes of climatic conditions. The solid construction of GEMA equipment, in which components that were sensitive to moisture were provided special protection, had proved itself splendidly. These lavishly constructed electronics cabinets, which were customarily made of cast aluminium, went into GEMA's very first series of production sets. Through their predominant ownership of shares in the

Rastatter Metallgiesserei (metal foundry) GEMA had gained direct influence over the production of their cast-aluminium chassis. Because the shortage of tools and material for moulds had become so acute, a new design of cabinets for the high-power transmitters was out of the question and construction had to be made using existing cabinets directly or with minor modifications.

It was difficult to make clear to the Navy and Luftwaffe that the introduction of new radars with appreciably greater range by means of high-power transmitters could take place only after considerable development and testing. Von Willisen, who was assigned to the operations of Naval Command West, was especially exposed to challenges because of the ever-increasing losses at sea. He undertook to introduce into operation in his sector much sooner than expected the new, powerful transmitters, some still laboratory models. The new power triode TS60 had in the meantime so matured that a small number of g-wave transmitters equipped with it could be produced. In order to gain time and appease the Navy, von Willisen allowed improvisation only on the modulator. His direct supervision of installations allowed deficiencies encountered to be quickly recognized and corrected. He was ably assisted by the unselfish enterprise of shop superintendent Henke, who demonstrated his value in every situation and at every location.

Since the Seetakt set Calais on the Channel coast had proved itself reliable, the Boulogne concept was tried out on it using various modulators yielding HF pulse powers of 150 and 400 kW on surface targets. The 150 kW transmitter Gisela and the 400 kW Grete were made available for this radar, although at a cost in accuracy. **The transmitters used anode modulation with spark switching of three capacitors that were charged to 36 kV by a cascade circuit, the capacitors being charged to 12 kV in parallel and discharged in series. Pulse length was determined by the duration of the spark and was adjusted to 2 μs by a load resistor. The spark gap was pre-ionized by a half-cycle of high voltage taken from the power net, leading to a spark and modulation frequency of 50 Hz.**

Using the finally developed Paris type radar, which used a stabilized, linear time-base triggered by the transmitter pulse, von Willisen installed the Paris–Boulogne with great success shortly before the Normandy invasion (1944). The first experiments with 150 kW power were encouraging. With the large Calais-B antenna he ranged ships at 40 km and tracked air formations to over 200 km. The range was only 30% greater than the Calais, but it had a much better capability to withstand jamming. In October 1944 the new set, designated FuMO5, was released for production.

It must be mentioned at this point that the Navy introduced new set designations for radar in 1944 that distinguished between sets for accurate location FuMO (Ortung, location), for general surveillance FuMB (Beobachtung, observation) and for electronic recognition and identification FuME (Erkennung, recognition). The first mobile Calais-A with 1000 Hz repetition rate and 1 kW peak power changed from FuMG39G(gB) to FuMO1; the subsequent Calais-B with 500 Hz and 8 kW changed from FuMG40G(gB) to FuMO2; the sets with bunkered electronics and rotatable hollow steel masts were designated FuMO3 and FuMO4 respectively. For ship-borne sets FuMO designations were initiated with two- and three-digit numbers, including as well special ground radars, the production of which GEMA was only partially involved in.

The well proven Messketten had to be replaced by linear time bases, which were triggered by transmitter pulses that were no longer related in phase to the master audio frequency because of the large scatter in the spark ignition time. Brandt set about finding a replacement and began with a design, suggested by his co-worker Wilhelmy, that allowed a restricted time span to be observed, locating single pulses accurately. The result of this work, which already led to a prototype in summer 1944, was a crowded adjunct for measuring range in existing sets having high-powered spark modulators. This pulse-shifted time base allowed the operator, as had the Messkette, to select a restricted span from anywhere in the 30 km total range so that the echo pulse was centred on the zero point of the CRT.

The radar on which this pulse-shifted time base, the modified OB110, was mounted acquired the designation Paris-2. It had a near range of 1–30 km and a survey range of 5–180 km wherein it could determine range to an accuracy of 1%. For the survey range 30–180 km there was an accessory that made 130–270 km possible. To simplify preliminary testing and calibration, the pulse-shifted time base for Paris-2 was constructed as a removable module. The cover of the console N was so altered that the Paris-2 could be inserted instead of the NB.

At the same time Paris-2 was being tested on the Channel the plant in Wahlstatt was testing a Freya-LZ outfitted with spark-gap-modulated transmitters having repetition rates of 50 and 100 Hz. It exhibited no deficiencies with respect to display quality, readability and reliability and brought strong approval from the Luftwaffe and the Navy. In demonstrations at Wahlstatt and at the front, the pulse-shifted time base proved capable of meeting the essential stringent requirements. Paris-2 allowed a significant reduction in the expense of range-measurement components of radars, as it replaced the audio oscillator, the Messkette, and the indicator that went with it. Because the circuit used two EC50 gas triodes, which were difficult to obtain, only 26 Paris-2 sets were built.

In order to have available range indicators for radars with spark-gap-modulated transmitters, GEMA restricted itself to the fabrication of 100 simple Paris indicators for coarse range estimates with calibrated screens (A scopes). The circuit of this device, which was modified from indicator OB110, needed the gas triode EC50. Those who did not experience the working conditions that existed at the end of 1944 will find it difficult to believe that this could limit production. The shortage also resulted in the Paris-2, which was only slightly less accurate than sets with Messkette, being deployed in very few numbers, and this excellent design remained nearly unused.

The spark-gap-modulated transmitters saw very little use by the Luftwaffe. The Freya-LZ that was modified for this kind of modulator produced great interest, as a 400 kW output with range of 250 km was dream-like for a mobile radar. Such powers allowed ranges that made the heavy, special sets Wassermann and

Mammut superfluous. The problem of suitable range-measuring circuits not being available led to the decision to use the standard transmitters for all mobile radars except for a few Freya-LZs held aside for experimentation. The Luftwaffe also showed little readiness to relinquish the Messkette and change to pulse repetition rates of 50 and 100 Hz. The objection to the low repetition rate resulted from the trace flicker, and GEMA thought they could overcome this through the use of long-persistence phosphors.

The antenna arrays of the long-range sets Wassermann and Mammut, as well as the panorama set Jagdschloss, were all changed without exception to common-mode operation, thereby using all elements for both transmission and reception. By mid-1943 GEMA had begun the production of a kit, DS, for altering to common-mode operation, designed originally for symmetrical feed. A high-emission thermionic diode, SD6/12/0.8, which had the same glass envelope as the decimetre triode TS6, was developed as a short circuit for the input of the receiver to replace the uncertain gas discharge. This diode would conduct when the transmitter voltage was impressed on it, but its impedance rose in the presence of small signals. By carefully locating the nodes of the transmitter's transmission lines, these tube characteristics combined to short-circuit the receiver input automatically. The original common-mode circuit with gas discharge had the disadvantage of the discharge impedance being large compared with characteristic impedance of the transmission line, which greatly reduced its effectiveness. With this new circuit it became possible to use normal low-impedance coaxial cables to feed the antenna asymmetrically. In order to use it on symmetric antennas GEMA made a matching unit, DWS.

To enhance the power of stationary air-warning sets they were equipped with anode modulation consisting of a power modulator with four TS41s and controlled by the TS module of the transmitter. The pulse voltage for the new transmitter 'Eibsee', which was introduced widely, was 20 kV, which allowed the output power of a transmitter with two TS41 tubes to reach 100 kW. This power modulator was so small that it could be fitted without problem into the old cabinets. Its weakness was in the transformer that raised the voltage of the modulator pulse. GEMA hoped that one of the high-voltage institutes that had research grants would find a solution to inhibit failures of this component, which resulted from insulation breakdown.

In order to bypass the critical and expensive pulse transformer, GEMA favoured the spark generator for anode modulation, which functioned with voltage multiplication and without the troublesome transformer. In this simple method capacitors, or a pulse-forming arrangement of delay lines, were charged in parallel and discharged in series through sparks. Two rows of spark gaps were connected back to back and the junction points received a pre-ionizing trigger pulse to start the high-voltage discharge. While these transmitter improvements were going on, Schultes had brought out a much improved f-wave receiver, codenamed 'Kreuzeck', based on work initiated at GEMA and carried to completion at Siemens. Its increased bandwidth allowed the pulse length to be significantly shortened. In combination with this new receiver some long-distance air-warning sets in North Germany and Denmark were equipped in early 1945 with laboratory transmitters, configured with double three-tube push–pull TS41s. With the spark modulator just mentioned it was hoped for a power of 500 kW, and there were reports of aircraft being observed over the east coast of England.

In January 1945 laboratory experiments at Wendenschloss Strasse used a Wassermann-L antenna and a transmitter that generated a 1.5 MW HF pulse with the 4 MW modulator Büffel. The newly developed TS100 was used in the transmitter Grete. Those who participated in these experiments remember fondly the spectacular fireworks that attended its operation.

The Navy deployed on the Channel and the Dutch and Scandinavian coasts 11 Paris–Boulogne sets with delay-line spark modulators Keiler and the large transmitter Gertrud, which had TS60s and yielded 400 kW. Nine more of these large sets were delivered, but all indications are that they did not see service. GEMA supported the NVK in a desperate situation in autumn 1944 in the development of a radar, codenamed 'Dünkirchen', for directing naval artillery with a new high-power transmitter and lobe switching. From this design one g-wave set with a 400 kW transmitter Gertrud and one c-wave set with a 1000 kW transmitter Grete were delivered in 1944 and early 1945 respectively.

In addition to this there was a series of activities by Dr Kober, who had replaced Schultes, that gave GEMA an absolute championship in range for a radar. The loyal co-operation with his predecessor brought him rewards not only in GEMA, for Siemens also gained through the collaboration, acquiring a foothold in radar through Schultes. Given their best efforts, GEMA was not in a position alone to bring the new series of Wassermann-M and Jagdschloss to maturity and to install them. (Jagdschloss was the Siemens model of the GEMA Panorama, which was tested on the tower at Tremmen. GEMA authorized Siemens to construct the tower and the rotating antenna for the series. The result was the Siemens Jagdschloss antenna with GEMA electronics.) Siemens became well known through this work for their long-range air-warning radars.

In mid-1943 Kober had the opportunity to renew experiments on a project he had undertaken earlier, a rotatable base for Freya that would make it a mobile panoramic set. The original investigations of 'Tremmen-Panorama' were carried out with a modified Freya to see if the presentation of these data on a plan position indicator (PPI) was practical with the phosphors of the CRTs then available. On returning to the work Kober used a high-power transmitter in common-mode operation with the antenna of a Freya-LZ to ascertain if target signals from 100 km were intense enough in the presence of noise and backscatter for intensity modulation of the CRT beam. The signal was passed through a filter that left only the peaks for brightening the electron beam in order to keep the screen from being made generally too bright. On leaving work on the Freya-LZ he considered another antenna support that would rotate at 10 revolutions per minute using a Wassermann rotary coupling. This auxiliary radar would present the operator with a continuous PPI picture of the air situation while allowing him to concentrate on specific targets with his main set.

But in 1944 there were no resources available at GEMA for this project. The Luftwaffe was interested in the mobile panoramic radar experiments but made GEMA turn this 'Dreh-Freya' (rotating Freya) over to Lorenz, who developed it as a PPI radar employing the GEMA Freya-LZ. It had a maximum range of 100 km. The PPI was located in a cabin separate from the rotating Freya, together with a modified electric power unit. The motor for driving the rotating base at a constant 5 rotations per minute could be switched from the cabin. The signals for the PPI and for controlling the antenna were transmitted through slip rings. The set could be used in modes other than circular sweeps, although it was not capable of presenting the PPI while examining individual targets separately, as Kober had originally intended. Only a few Dreh-Freyas came into service. They had the tube complement for high-power modulators and complete Wismar anti-jamming circuits. The broad-band antenna, codenamed 'Alpspitze' (alpine peak) had four arrays of four horizontally polarized dipoles divided into two segments.

Because of the threats to which radar stations were subjected, the Luftwaffe required that the electronics be placed in bunkers. For some time the Navy had placed the Calais radars with their electronics and antenna, which was retracted into its lower idle position, inside concrete rings that provided protection from the bursts of bombs or shells striking nearby, as well as making the positions more difficult to observe from the air. For ship-borne sets that were not mounted on optical rangefinders or on special radar mountings, the firm of Beeck in Oldenburg was favoured for providing rotatable hollow-steel masts for antennas. The size of this so-called 'Zerstörersäule' (destroyer column), which was placed on the bridge or otherwise on the deck, varied according to the size of the vessel. The electronics were located wherever it was convenient, usually in the radio room.

In summer 1944 GEMA ordered from Beeck rotatable masts to be used for the broad-band antennas of fortified Freya-LZ stations in a manner similar to the Calais sets that were being placed along the Atlantic Wall. The operating crew and the electronics were to be protected from enemy fire by being housed in a bunker or dugout. The sets received were designated Freya-LZ verbunkert (fortified). There was room neither in the facilities in Berlin nor in Lauban to assemble and test such equipment, and in order to avoid superfluous use of vexatious transportation, GEMA rented in Oldenburg 10 000 square metres of grounds and 4000 square metres of a former wagon shop for assembling and testing these special radars.

It came as a surprise to GEMA to learn that their special sonar Mime for U-boats had seen no important action by the Navy. They had recognized the improvement in locating capabilities but discontinued series production because of the mechanical expense. For a considerable time it had become obvious that there was little fondness for the SH and provisional SZ sonars because of the half-silvered mirror CRT indicators. Mime had offered a definite improvement in range measurement, but the mechanically moving mirrors were disliked, and since 1943 the Navy had insisted vigorously on a method that displayed the data entirely on a CRT, as was the case for radar indicators.

Ever since these complaints had been heard, GEMA had undertaken appropriate investigations aimed at correction in a device codenamed 'SR-Kino'. Much earlier, indeed in January 1941, GEMA had begun a preliminary development and construction of the G/H model together with an adjunct set with electric time base and a smaller twin amplifier. Its limits for range were 50–210, 150–630, 500–2100 and 1500–6300 m. The first experiments were carried out with CRT DG16 and LB7/15 but were not satisfactory because the phosphor persistence of these tubes was too short. In 1942 GEMA ordered from Opta-Radio in Berlin a special long-persistence CRT with a flat, rectangular screen 180 by 80 mm. The development of the indicator G/H using this new CRT was delayed by the design of a new twin amplifier with Wehrmacht tubes. The principal difficulty lay in designing the extremely important amplifier range-related gain control.

Because the amplitude of a sound pulse in water decreases rapidly with distance, it had been shown experimentally to be advantageous to have the amplifier gain increase as a function of the time elapsed from the emission of the sound generator pulse. The difficulty lay in the high degree of equality of gain and phase for the twin amplifiers that had to be maintained over the control region in order for the sum–difference method to function. Since the solution for travel-time gain control for the new small twin amplifiers was at hand, the Navy wanted the regular twin amplifiers to have gain control incorporated. These modified twin amplifiers were to be delivered, however, with the SZ and provisional SH sonars.

The new, flat CRT for the electronic sonar indicator could not be obtained early enough for series production, so GEMA chose a simpler and faster way to build the indicator. In their first radars they had used a vertically standing single-beam CRT of type DG16 that was observed with a mirror. This design now found rebirth. Earlier it had been dismissed when the AEG dual-beam tubes were introduced. **It introduced the concept of an indicator that used the 180 mm diameter CRT type HR 1/180/2 that was used for all range measurement in the next sonars and reminded one of the first radar sets. The new indicators, which now, on the explicit demand of the Navy, had to be outfitted with steel vacuum tubes for the first sonars 'Nibelung 1A', type SG 1001, and 'Nibelung 1B', type SG 1101, contained an acoustic adjunct SGD 101, a twin amplifier SGZ 101, an indicator SGK 101 and a command transmission BÜ-attachment SGÜ 111.**

A previously developed unit for the acoustic adjunct was used in which the frequencies 10 or 15 kHz were converted by heterodyne to the audible 2 kHz, generating small differences in frequency resulting from the Doppler effect that allowed the motion of the target to be determined. The previously developed twin-amplifier unit A/B was also provided with heterodyne detection.

Another petty behaviour of the Navy came about in connection with the development of the new GH sonar indicator with a linear CRT time base. They ignored

GEMA's development of an electronic indicator because the twin amplifiers, audio amplifier and sweep circuit used Luftwaffe tubes, which were much more expensive than steel tubes. Range-measuring devices equipped with such tubes had been extensively tested and put into production in summer 1944. Altering the design to steel tubes cost an extra four months, so that by the end of the war very few Nibelung sets had been tested and put into service. Parallel to this set was another carrying the codename 'Notung'. It differed from Nibelung in the BU attachment (command transmission). GEMA fabricated a number of experimental models in which curiously only Notung I was outfitted with steel tubes. In the next edition, Notung II, the earlier twin amplifier with integrated audio amplifier with Wehrmacht tubes was used.

GEMA's low-frequency laboratory was able to carry one interesting idea through to prototype in 1944: 'Undine', UT-Boje (buoy), an underwater sound telegraph that made it possible or helped submerged U-boats to gain entrance to narrow waters, especially fjords. It was a relay transmitter contained within the shell of a mine. This buoy had an underwater sound function similar to what the IFF Erstling had for radar. It was powered by storage batteries that provided energy for about two months' operation. The buoy contained a transducer for transmission and reception, a receiver amplifier with code decryptor and a sound transmitter with coded answer.

The activation of the buoy came about through a series of coded sound pulses sent by the U-boat's sonar. The basis chosen for the code was the number of pulses relative to the time between them. On recognizing the call the buoy switched on its transmitter and sent a series of coded answer pulses, which could be received by the U-boat's sonar and the direction ascertained.

The transducer consisted of multiple strips, arranged to generate narrow sound radiation in a vertical dimension. The receiver had a tuned amplifier with space-charge pentodes RV2, 4P45, designed for an anode voltage of 24 V. The interpretation and decryption of the pulses received was accomplished with a circuit that used special, sensitive relays. The operation of the transmitter had similarities with that of the FuG25A Erstling, a self-excited transmitter using the LS50. The anode potential for the transmitter used the ac–dc converter and power supply from Erstling to provide 1000 V. The output sound pulse power was 500 W and had a 5 μs pulse length. The encoding device for the transmitter was also taken from Erstling. Two substitutable keys varied the rhythm and pitch of the transmitted pulses, which allowed them to be recognized as a melody.

One of Brandt's last inventions, which he initiated in September 1944, demonstrated how GEMA was able to continue with the development of new ideas into the last months of the war and the tempo with which, despite all hindrances, they created new experimental models.

The reliability of the anode-modulated high-power radar transmitter that used spark gaps as switches had become so great as to point toward its use as a high-power sound generator for sonars. This offered not only much higher powers but would also require less space and weight, and the small effort required justified the project development and testing. In August 1944 Brandt presented the idea of building the sound generator on the basis of experience with spark modulators to the naval scientific leadership and a month later obtained an order to develop a sound generator and to build an experimental model. This, GEMA's last big development project, carried the codename 'Laurin'.

Laurin was divided into power supply, modulator and transmitter. The transmitter was modulated by a delay line that had been charged from a high-voltage transformer through rectifier VH3 and that furnished 20 ms rectangular 5 kV pulses. This short-duration, constant potential yielded good efficiency and constant frequency for the TS41 oscillator, which gave 6 kW at a pulse repetition period of 3 s. The modulation used a pre-ionized gap, the technique known from radar, and employed two series-connected spark gaps. This construction composed a compact unit only 500 mm wide, 350 mm high and 400 mm deep, substantially smaller than the AEG generator, and its weight was only 70 kg. On 19 January 1945 a laboratory model of Laurin was sent for testing to the industrial test section of the NVK.

GEMA concluded the business year 1944 admirably. It shows the high degree of willingness of its 6000 employees, male and female, in night and day shifts under difficult war conditions that allowed the company to fulfil the high demands made of it by the government. The preceding year had been distinguished by a turnover just under 100 million RM ($24 million): this increased to 131 million RM in 1944, which was all the more impressive because the large amount of series production brought with it reductions in price, hiding an even larger production of goods. In 1944 302 sonars of various types and 1206 radars were delivered, of which 25% were from subcontractors AEG, Telefunken and Lorenz. In addition GEMA delivered to Siemens the electronics for 148 Wassermann and Jagdschloss sets. A non-negligible part of their output was about 10 000 Erstling IFF sets for the Luftwaffe, of which 68% could be credited to GEMA's production. Much of this equipment never saw service because it was destroyed or damaged either in transit or at the destination.

GEMA's production accomplishments were often unauthorized, even objected to by the procurement authorities or by the radar working committee. The reason lay in the vexation of certain persons, who made such remarks as 'GEMA took over completely' or 'one cannot get anything out of GEMA' or 'GEMA never meets a deadline'. In addition there was discrimination in the form that 'GEMA's products are not reliable and are technically backward'. It was in fact extremely difficult to be right when dealing with the Navy and the Luftwaffe, and it was not easy for the two managers to keep above the invariable competition between the two services. A subsequent examination of the debit and credit sides of the deliveries by their firm lets one see, however, that both the Navy and the Luftwaffe had sufficient radars at their disposal for the tasks facing them. If they had controlled the deployment of this equipment with more concern, and if they had organized its operation promptly, then the deficiencies that were credited to GEMA would not have arisen.

The Calais earned the reputation as a most reliable ground radar, along with Freya. Its hasty introduction soon determined that sets with higher capabilities were called for, but the older GEMA sets were not to be beaten, because the consequent evaluation of failures had removed their teething troubles. The robust construction and clean engineering had proved itself worth the large material outlay, and this applied equally to sea-borne Seetakt sets. They were exposed to climatic and combat conditions that necessarily led to more failures than ground radar. The number of radar failures that were repaired and returned to service through replaced modules or direct repair went down markedly with the increase in the quality of the training of radar personnel aboard ship. Here the robust construction and clean engineering proved effective.

Until the 1944 invasion the Navy alone used about 2000 GEMA radars of all types aboard ship and on land. By February 1945 6000 units had been manufactured for the Navy and the Luftwaffe. The number of sonars is less because the war at sea shrank in extent with time. GEMA had produced a large part of the active underwater sound- and radio-location equipment. War determined that GEMA became an important supplier of the Wehrmacht, of course. One has to assign responsibility to the regime of those years that no equipment was ever developed or manufactured for civilian use, a consequence of the exorbitant secrecy that the Navy imposed at the start.

CHAPTER 20

GEMA'S CONSTRAINED ENDING

By the beginning of 1945 the war situation for a battered Germany made it obvious that an evil end was at hand. On 12 January the Russian offensive at the centre of the East Front began, and in short order Russian troops were standing in Upper Silesia and on the Oder. With that there began for Erbslöh and von Willisen a time for difficult decisions that would determine the company's future. Just to save the portions of the plant that had been moved to the east, they had to deal with fools who still believed in final victory and who would not hear of a withdrawal.

When the extent of the Russian offensive became known von Willisen carried out a trick that allowed him to begin moving the Wahlstadt plant. He would first assert that he wanted to shift the laboratory and construction work to Lauban, when in reality he was planning to move the whole plant to Berlin-Köpenick by way of Dresden. Brandt was given responsibility for the operation. Late on the evening of 13 February he started loading buses and trucks with a part of the laboratory personnel together with their personal possessions and what laboratory equipment had not been sent in advance. The night trip through Görlitz, Löbau and Bautzen was halted at daybreak of 14 February by tyre trouble in the vicinity of Dresden, which the RAF had completely destroyed in two attacks during the previous night. This caused further direct movement towards Berlin to be interrupted and the caravan quartered itself temporarily in various localities, such as the Sachsenwerk in Radeberg, among others.

With the foresight that Berlin was no longer a safe place either and would presumably fall into Russian hands, von Willisen had already selected Lensahn in Holstein as the final location for the laboratory, and for the construction and fabrication of prototypes, to which he now directed his column. He had already picked a dance hall for the continuation of laboratory work and a government granary for the storage of supplies. The Navy made some large barracks available to him at Pelzerhaken. The further work demanded by Naval Command continued only in small amounts because electric current was rationed. Besides the development projects, such as 'UT-Boje' and 'Laurin', work was proceeding on radar methods of fighter control using the IFF set Erstling as well as on the intended development of 'Egon' and 'Wachtel'. In the field of underwater sound there was a passive locating technique with the name 'Felchen' under study that would determine the direction and range of sound-producing targets. Following urgent needs professed by the NVK, GEMA worked on oscillators, sound generators and receivers for frequencies around 10 Hz to be used for measurements of ocean currents, devices carrying the codenames 'Steinbach' and 'Steingraf'. On 26 April 1945 Naval Command, now located at Eutin, ordered GEMA to have their refugee crews in Lensahn and Pelzerhaken carry out repair of radar and sonar equipment in addition to the remaining development tasks.

Production in the Lauban plant had in part to be continued until 17 February, a Saturday morning, although since 6 February machines and material had been loaded on railway cars and dispatched to various refugee locations. Suddenly, at 9:00 AM, in the midst of an inspection of the plant an artillery attack began. The first volley struck Plant No 1, and the administration building received direct hits. Without further delay Plant Manager Golde organized the evacuation of the personnel and conducted a prepared destruction of the more important parts of Plants No 1 and 2 as well as the outlying rooms that contained supplies that had not been evacuated. Scattered soldiers from General Wenck's troops and members of the Volksturm (home guard) were able to stop the Russian advance briefly, so that Golde and a handful of courageous workers were able to enter Plant No 2 again and extract more machines and tools so long as the time allowed.

The remaining members of the group agreed that the retreat, which had now become a flight, was to proceed over Schönberg to Warnsdorf. Actually nearly 400 co-workers came to Warnsdorf by motor vehicle, bicycle and foot. Golde ordered about 200 of them to proceed to the main plant in Köpenick by way of Tetschen-Bodenbach and Bischofswerda. He divided the others between three locations: Waldsassen near Eger, Falkenstein in the Vogtland and Weferlingen near Braunschweig. It is an example of Golde's sense of duty in February 1945 that he believed he would be able to resume production in Waldsassen. Because the war situation at the main plant in Berlin-Köpenick had brought production there nearly to a standstill, he stopped movement of some of the cars loaded with machines and material and continued to Berlin only with the completed sets they carried with them. In despair Golde attempted underway to gain some kind of communication with the scattered members of his little band, either to get them settled somewhere or to bring them to Waldsassen. At the beginning of March Naval Command assigned that part of Golde's wanderers from Lauban some room with the firm Gareis in Waldsassen.

The railway cars with machines and material had been held up at Fulda, which gave Naval Command the idea that GEMA could set up a plant there for a short time and supported GEMA in their search for suitable space. This search soon brought together nearly every possible governmental authority just to gain some idea of what might be available. Every undestroyed bit of space was or would be occupied by refugees from the east. An extract from the travel report of a GEMA employee charged with locating space in Fulda will quickly show the difficulties experienced in travelling and even in just existing:

'···after taking care of the travel papers I was at the Anhalter Station on Wednesday, 21 February at 16:00 hours in order to go to Fulda. It was only possible to travel by train at night from the Potsdam Station to Magdeburg. From the Magdeburg main station I went by foot to Magdeburg-Buckau in order to reach Fulda by stages using local trains passing Halle, Weimar, Erfurt and Bebra, arriving in Fulda on the 23rd at 20:00 hours.' (The distance from Berlin to Fulda is approximately 250 km.)

All of Golde's attempts to reconstitute a working factory from the scattered parts of the two Lauban plants shattered in the realities of the situation. Many cars were bombed or shot up by low-level fighter attacks while others were simply lost. The war made any reasonable operation impossible.

On 16 April the Russian offensive began on the Oder with the goal of surrounding and capturing Berlin; on 25 April the ring was closed after Köpenick had already been reached by Russian troops. Until that moment the activity at the main plant had been to prevent important papers and equipment from being taken by the enemy, but that did not succeed and really made no sense. Countless GEMA workers fled out of Berlin to escape the Russians and did not experience the occupation of the works on Wendenschloss and Achenbach Strassen.

Brandt with a group of about 120 was able to break through to Lensahn in an adventurous manner and attempted to set up his laboratory there. In March a control office, called 'Stützpunkt Blücher', had been set up in the Berlin plant to co-ordinate and watch over GEMA's scattered transport, but it became impossible to keep informed of where on rail, road or water the disembodied company existed. Bits and pieces of a laboratory—boxes containing experimental sets, supplies and instruments—continued to arrive at Lensahn until mid-April, and Brandt still entertained the idea of reopening some kind of laboratory. After the hypothetical outfitting of this future laboratory had been established, for which a source of electric current proved to be the greatest problem, his plans had to face the capitulation and occupation by British soldiers.

On 23 April the Red Army was in Köpenick and had taken possession of all GEMA property. After a long and thorough dismantling the Soviet Military Administration created a scientific-technical institute, the MSP, in the intact buildings on Wendenschloss Strasse in which a number of specialists from GEMA, Siemens and Askania employed their knowledge to the benefit of the new rulers.

In fall 1949 the military authorities vacated the buildings of the former GEMA works in Berlin-Köpenick and transferred them in an agreement on 1 October 1949 made with the state-run factory RFT of the German Democratic Republic that declared them public property. In 1946 the mayor of Köpenick appointed an economic examiner as trustee and executor of GEMA's remaining assets. He attempted until 1949 to locate the scattered objects and supplies in all the Zones of the occupying powers but was able to accomplish this only to a limited extent.

In 1949 two new factories were founded as GEMA's successors: the VEB (State Industry) RFT, a central laboratory of communications and special equipment, and the VEB Funkwerk Köpenick. In these two concerns GEMA had worthy successors. In their fabrication programme was equipment for electronic instruments, broadcast stations, maritime communication and electronic navigation, gyroscopic apparatus, underwater sound and ship control.

On 31 May 1945 GEMA at Lensahn was officially closed. This happened after the British authorities had thoroughly cleared out the German war material there so they could use the space to intern Wehrmacht personnel being returned from Norway and Denmark. This clearance took place on 16/17 May, and all of the tirelessly saved experimental models and prototypes of GEMA development were destroyed, so that when the research-minded British Intelligence Service arrived, only wreckage remained. With capitulation the Military Government had forbidden von Willisen from further operation of GEMA. In order to find work for some of his co-workers he founded a new firm with the name 'Mechanische Werkstätten Lensahn'. The authorities allowed them to repair agricultural machines and equipment. Later they expanded permission little by little to allow the manufacture, as well as the repair, of radios. Out of this grew a small factory for making condenser microphones, amplifiers and loudspeakers housed in an old RAD (National Labour Service) barracks and in the rooms of the pub 'Grüner Hirsch' in Oldenburg/Holstein. The company grew until 1947 under the altered name 'WILAG, Willisen Apparatebau' to nearly 400 workers. In competition with many companies in the field it had a difficult time and had to close in autumn 1948.

CHAPTER 21

TRANSLATOR'S EPILOGUE

In 1939 eight countries were independently and secretly developing radar. All started in the mid-1930s, the consequence of the availability of two key electronic components: low-voltage, high-vacuum cathode-ray tubes and multi-grid amplifier tubes. Three of these nations quickly out-paced the others and were soon in a technological race without realizing it, as each thought radar to be exclusively its own. It is useful to compare briefly GEMA's early achievements, so carefully and accurately explained in these pages, with those of Britain and America. How were they similar, and how did they differ?

GEMA's efforts find a remarkable parallel with those of two American laboratories: the Naval Research Laboratory (NRL) and the Signal Corps Laboratory. Both had given thought to radio location at the beginning of the decade, and both had initially sought a solution through microwaves. The Signal Corps conducted some experiments with Irving Wolff of RCA, who used Barkhausen tubes with paraboloid reflectors. The experiments were able to measure the speed of automobiles through the Doppler effect but failed at anything approaching radio location. The Director of the Signal Corps Laboratory, Major William Blair, pushed this approach, until it was clear that this path would lead nowhere until a powerful generator of centimetric waves became available.

NRL and the Signal Corps found, as did GEMA, that arrays of metre-wave dipoles produced a narrow beam, which if not the ideal 'radio searchlight' that had been sought, showed immediate and significant promise. Initially the Signal Corps wanted a device for pointing searchlights at aircraft, the SCR-268; NRL something that could protect warships from surprise air attack, the CXAM. GEMA wanted to detect and range surface vessels accurately.

All three laboratories were remarkably similar in their approach to the technical problems presented. All used high-gain antennas and pushed output triodes to the limit to attain powers at levels of tens of kW. In all three the work was initiated by engineers in the laboratory, indeed at their work benches. The two American labs used funds intended for other projects, and GEMA used private capital earned by the two entrepreneurs, Erbslöh and von Willisen. All three secured government funding specified for radio location only after they had demonstrated a successful product. The impetus for all three was the realization that such a device was possible and was judged by the engineers to have military value.

In the case of Britain the opposite situation held. Germany's rearmament caused concern in the Air Ministry, which inquired of the Radio Research Station at Slough if there were methods useful for air defence. The inquiry was answered by Robert Watson Watt and Arnold Wilkins, who quickly devised a method of radio detection based on their own experience, which was ionospheric sounding with waves of tens of metres in wavelength and locating distant lightning discharges. They were given what amounted to a blank cheque and found the enthusiastic support of Hugh Dowding, one of the few high-ranking officers of the Royal Air Force who thought a creditable defence could be made against bombers with fighter planes. The British effort became large rather quickly and provided an effective air defence by 1939, whereas the American and German work only produced good designs ready for or in limited production. The British development also differed from the American and German in that it did not make use of communication engineers until the project was well under way, as Watt thought engineers were too narrow in their approach. Although Britain had an outstanding electronics industry, it was not approached for design work. Watt's technique used broadcast transmission operating on tens of metres and at hundreds of kW; the direction of the reflected signal was obtained by comparing the amplitudes obtained from crossed dipoles. In 1939 W A S Butement and P E Pollard, engineers for the British Army, built a radar for coastal artillery with design elements very much the same as those used by GEMA and the two American laboratories. It was soon used to correct deficiencies in Watt's long-wave system.

Before leaving these comparisons it is worth pointing out another German–American similarity: the nearly parallel work done by Telefunken and the Bell Telephone Laboratories. Although Wilhelm Runge of Telefunken brusquely dismissed Kühnhold's early inquiries about radio location, creating thereby an enduring enmity, he had second thoughts later and invented the justly famous Würzburg gun-laying radar. NRL's approach to industry was not rebuffed, as Bell responded quite favourably and began developing what became the Navy's FD (Mark 4), an electronic cousin of the Würzburg, both working in the decimetre band, with its own funds. The Würzburg was the Luftwaffe's principal antiaircraft radar throughout the war; the FD was the US Navy's dual-purpose ship-borne radar throughout the war. The Royal Navy, employing engineers independently of the Air Ministry at His Majesty's Signal School, followed the same approach, developing the types 282 and 261, very successful dual-purpose ship-borne radars.

Thus we see that radar was something that was very much 'in the air' at the time. When faced with the need or taken with inspiration, engineers and physicists applied the electronic knowledge that filled their handbooks and produced radar on demand. Under these conditions matters of priority have only minor significance, although the GEMA team can justly claim to have first produced what can rightly be called a radar in their demonstration on 26 September 1935 of a set that determined direction and range accurately enough for blind naval fire—only to see this outstanding capability ignored for something simpler to operate!

An aspect of the German effort that seems to have differed from the Allied was the degree to which corporate rivalry affected the course of events. The numerous agreements that had to be made concerning licensing and post-war rights in order to smooth production will certainly seem remarkable to American and British readers. This may have come about because radar research was carried out to a great extent in government laboratories in America and Britain with a smaller contribution from corporations. A puzzling aspect of German radar research was the delay imposed by severe secrecy in drawing on the many excellent universities and polytechnic institutes until very late in the war, whereas America and Britain made full use of these faculties from the very beginning.

SOURCES

1 'Geschichte der deutschen Funkmesstechnik', tape recording of an account by Hans Karl Freiherrn von Willisen, 1952.

2 'Ein Teil meines Lebenslaufs', Paul-Günther Erbslöh to the author, 1990.

3 Letter of 1 December 1950 to Erbslöh with an outline for a GEMA history ('Die Entwicklung des deutschen Ortungswesens-Radar') by Dr Fritz Walter, Augsburg, former patent attorney for GEMA.

4 'Erinnerungen an die GEMA 1935–45', unpublished manuscript of Dr Walter Brandt, Bochum, 1986.

5 Summation of interviews of Dr Rudolf Kühnhold imparted to Herrn Vanderhulst, 1985. Obtained from Claus Kinder, Kiel, 1987.

6 H Frühauf, 'H.E. Hollmann zum 60. Geburtstag', Hochfrequenztechnik und Elektroakustik, vol. 68, pp. 141–143, December 1959.

7 Leo Brandt, Zur Geschichte der Radartechnik in Deutschland und Gross Britannien. Genoa: Pubblicazioni dell'Istituto Internazionale dell Comunicazioni, 1967.

8 Herbert Grenzebach, Ein Leben für die Telefunken-Schallplatte. Düsseldorf: Hansfried Sieben, 1991.

9 'Arbeiten auf dem ULTRAROT-Gebiet', report of Dr Grosskurth (NVK) to the Allied Control Commission, 15 July 1945.

10 Account of H K von Willisen concerning the last voyage of the research boat *Welle*, 15–20 January 1937.

11 Hans Bertele, Einführung in die elektrischen Impulstechniken, vol. 1. Wien: Oldenbourg, 1974.

12 Attested copy from the Handelsregister, Amtsgericht Berlin concerning GEMA, 10 December 1941.

13 Abstract of the agreement of affiliation, 16 January 1934, among the members of the firm GEMA, Berlin-Wilmersdorf between Paul-Günther Erbslöh, Georg Neumann, Erich Rickmann, Hans Karl Frh von Willisen, the firm Tonographie, GmbH and the firm Georg Neumann and Co.

14 Summary of GEMA's principal patents as of July 1935 by Dr Fritz Walter, Patent Attorney.

15 Supplementary list of GEMA's principal patents, August to December 1935 by Dr Fritz Walter, Patent Attorney.

16 GEMA-Patent No 729 831, Reichs Patent Office, 15 December 1933, 'Lotungs- und Entfernungsmessverfahren mittels reflektierbarer Wellen'.

17 Document 1144 of the GEMA patent application of 27 April 1935 concerning 'Ein Verfahren zur Entfernungs- und Richtungsbestimmung von Gegenständen mittels reflektierter elektrischer oder akustischer Wellen'. Dr Fritz Walter, Patent Attorney.

18 'Beschreibung der Funk-Fernsprech-Anlage Type FA75'. GEMA, 1937.

19 'Beschreibung der Funk-Fernsprech-Anlage Type FA80'. GEMA, 1937.

20 S-M-Anlagen, circuit diagrams, GEMA-Schaltung 607, 4 July 1939.

21 Anlage Typ S173A, circuit diagrams, GEMA-Schaltung 607, 27 October 1939.

22 Schedule of terms for GEMA sonars, 19 September 1939.

23 'Ermittlung der Verwendung der vom OKM der GEMA zur Verfügung gestellten Mittel', Management Examiner Emma Adolph, 1939.

24 Carbon copy of the contract between the German Reich (Wehrmachtfiskus), represented by the Commander of the Navy, and the firm GEMA, effective 1 January 1937.

25 Contract between the firm C Lorenz, Berlin-Tempelhof, and the firm GEMA, Berlin, with appendix listing GEMA's patents, 30 December 1935.

26 Agreement of understanding between the firm C Lorenz and the Naval Command concerning the dissolution of the contract between Lorenz and GEMA, 23 October 1936.

27 Survey of the DeTe equipment as of 23 March 1939.

28 'Geheimer Aktenvermerk über Kollisions-Schutz- und Lotsen-Gerät für Friedensfertigung', GEMA, Golde, 6 August 1940.

29 'Kurzbeschreibung des Magnet-Kinos, Type SGK168', from a report of type acceptance test of 6 September 1940.

30 Notes concerning type acceptance test of the Magnet-Kino for sonar type SGK168, 6 September 1940 at GEMA.

31 Notes concerning the development of the shop truck to repair radar sets in the field, 26 March 1940.

32 'Beschreibung der S-Anlage für U-Boote, Type SH104, SH106 und SH113, GEMA Type S203', 19 September 1941.

33 'Beschreibung der S-Anlage für neue M-Boote, Type SF700 bis SF702, GEMA Type S186A, S186E und S186A/1', 19 September 1941.

34 'Betriebsanweisung für S-Anlage mit Handausfahrgerät', edition of October 1941.

35 'Beschreibung der S-Anlage für U-Boote, Type SH205, SH206 und SH208, GEMA Type S206', 19 September 1939.

36 'Abkommen zwischen der GEMA und Telefunken über den Abschluss von Verträgen zur Weiterarbeit auf dem Gebiet der Funkmessgeräte nach dem Kriege', 11 December 1941.

37 'Abkommen zwischen der GEMA und der Allgemeinen Elektrizitätsgesellschaft über den Abschluss von Verträgen zur Weiterarbeit auf dem Gebiet der Funkmessgeräte nach dem Kriege', 20 January 1942.

38 'Abkommen zwischen der GEMA und Lorenz AG über den Abschluss von Verträgen zur Weiterarbeit auf dem Gebiet der Funkmessgeräte (Ortung terrestrischer Ziele und Ortung z.B. von Flugzeugen, Schiffen, Meeresuntiefen, Eisbergen)', 28 April 1942.

39 'Erläuterungen zur Typenbezeichnung der S-Anlagen', GU56-00104, edition of January 1942, Kriegsmarinewerft Kiel, Nachrichtenmittelressort.

40 Werkschrift 3014/1 (GEMA), 'Gerätehandbuch und Beschreibung Dete-Gerät II, FuMG(Flum)40G(fB)', March 1942.

41 'Beschreibung für FuMG(Seetakt)40G(gL-gS), GEMA/OKM', edition of April 1942.

42 'Nachtrag zur Beschreibung für FuMG(Seetakt)40G(gL-gS)', GEMA/OKM, edition of February 1943.

43 Abstract of the guidelines for an agreement between GEMA and Siemens, 30 October 1942.

44 Contract between GEMA and Siemens & Halske concerning collaboration in the field of radar, retroactive to 1 October 1942.

45 GEMA production report for 1943.

46 License agreement between the firm Atlas Werke, Bremen, and GEMA, Berlin-Köpenick, dealing with claims arising out of patent 470 782, 27 August 1942.

47 Judgement of the 1st Infringement Court of the Reichs Patent Office in the case of the firm GEMA against the firm Atlas Werke concerning the declaration of ineffectiveness of patent 470 782, 25 March 1943.

48 Extract from 'Ausbildungsvorschrift für die Flakartillerie', Heimatflak, 25 February 1943.

49 List of names of the GEMA workers pressed into service in an antiaircraft artillery and a searchlight platoon for the defence of the plant from 3 November 1943 to 27 July 1944.

50 Abstract of a contract between GEMA and Herrn Marinebaudirektor Dr Rudolf Kühnhold, Kiel, 1 January 1944.

51 Guidelines for a GEMA contract with the physics departments of universities and technical institutes, 28 June 1944.

52 'Vorläufige Beschreibung und Schaltanweisung zur Erfassung des Nahbereiches bei S-Anlagen', GEMA edition of February 1943.

53 'Beschreibung der S-Anlage für Oberwasserschiffe, Type SH400b bis SH414b, GEMA Type S195A/1, S195E/1, S197A/1 und S197E/1', 1st edition, February 1944.

54 'Funkmessgerätekunde', Oberkommando der Kriegsmarine, 1st edition, February 1944.

55 'Lehrunterlage "Dreh-Freya"', Oberkommando der Luftwaffe, Generalnachrichtenführer, September 1944.

56 Naval technical manuals: Na 235590, test transmitter 'Enzian f' for f-wave; Na 235592, test transmitter 'g' for g-waves; Na 235553, test unit II, type PG104; Na 235550, test unit PG101, Na xxxxxx, test unit PTN1, Na xxxxxx, test bench K, PTK2.

57 Technical manual No 346, radar FuMG(Flum)40G(fB).

58 Technical manual No 89, GEMA radars.

59 Technical manual for radar FuMG401(Freya), February 1945.

60 TM E11-219, Directory of German Radar Equipment. Washington: US Government Printing Office, April 1945.

Figure 1: Dr Rudolf Kühnhold, scientific leader of Nachrichtenmittel-Versuchskommando (NVA/NVK), the research laboratory of the German Navy, shown here as Baudirektor der Kriegsmarine (director of naval construction). Born 1903, died 1992.

Figure 2: Paul-Günther Erbslöh, co-founder and business director of GEMA. Born 1905.

Figure 3: Hans-Karl Freiherr von Willisen, co-founder and business director of GEMA, shown here in naval uniform as consultant and instructor for radar. Born 1906, died 1966.

Figure 4: An early Tonographie production installation for the transfer from sound film to wax disc phonograph recordings, which were required by cinema houses still having silent projectors. The records were 40 cm diameter, played at 33 1/3 revolutions per minute and were cut from the inside out. 1931.

Figure 5: Von Willisen and Erbslöh seen engaged in measuring the range of 48 cm waves at Bismarck Heights near Potsdam. Behind them is seen a broadcast receiver, Bayreuth, which was used as an IF amplifier. 1934.

Figure 6: The first 'GEMA tower' on the NVA grounds at Pelzerhaken. 1934.

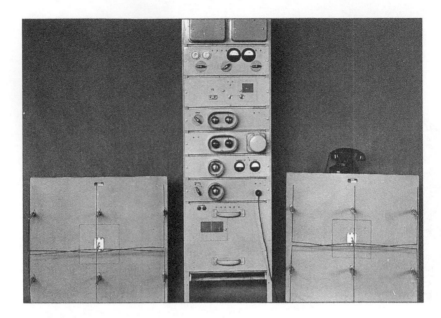

Figure 7: GEMA radio telephone FA75 for 75 cm wavelength: front view. To ensure security of transmissions, this equipment incorporated speech scrambling, if security of transmission was required. 1936.

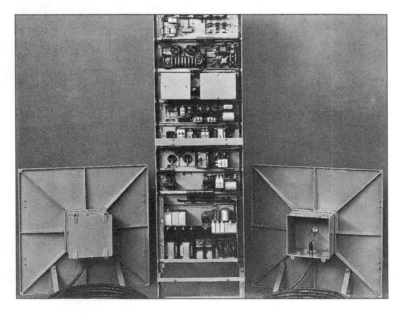

Figure 8: GEMA radio telephone FA75 for 75 cm wavelength: rear view.

Figure 9: NVA research boat *Grille*. Because Hitler's yacht carried the same name, the vessel later became *Welle*. 1934.

Figure 10: The grounds of the naval arsenal at Kiel around 1945. In the centre is the BG tower on which a Seetakt antenna can be seen.

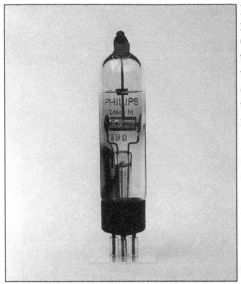

Figure 11: The Philips magnetron (without magnet) with which GEMA developed the first 48 cm transmitter with sufficient power for use in radar. Although soon abandoned for high-frequency triodes, this tube created much enthusiasm for radar at the time. 1934.

Figure 12: The transmitter unit Ultrateil. Two high-frequency GEMA triodes, TS1 and its mirror image TS1A, were mounted for push–pull operation in a resonant assembly that generated 60 cm waves at 1 kW peak power in 1935. This historically significant element was the basis for the design of all subsequent transmitter outputs. The photograph illustrates GEMA's modular approach to construction.

172

Figure 13: During 1934/35 GEMA used the American type 304 tube as the basis for the design of their own TS2. It had little value for decimetre waves.

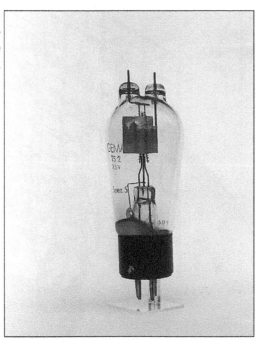

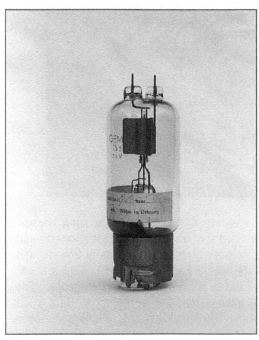

Figure 14: The TS2 evolved into the TS5 in 1936. It had a bayonet socket and was primarily used in sonar equipment.

Figure 15: An early version of the f-wave (2.4 m) radar that soon became known as Freya. The design was intended for field operation that allowed rapid deployment. For transport the antenna assembly was hydraulically lowered and dismantled. With attached wheels the whole unit was pulled behind a truck. This resulted in it being called the Limber Freya by British intelligence. The transmitter used the lower half of the antenna, the receiver the upper. Hard wood was employed for the antenna frame and the base at that time. Photograph taken on Sehberg near Kiel in early 1940.

174

Figure 16: Early series production of the Limber Freya, shown with an engine-driven generator as a power source. 1940.

Figures 17 and 18: Construction of a Freya from an air-transportable package deployed on the Norwegian coast during the invasion. April 1940.

Figure 19: With this TS4 GEMA attained ranges exceeding 100 km using the 2 m wavelength band.

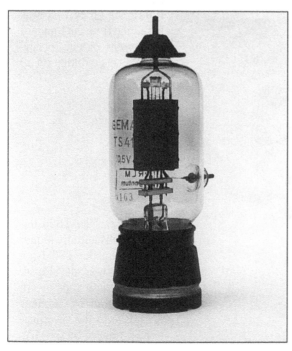

Figure 20: The TS41, a further development of the TS4, became a general transmitter workhorse. With six tubes in parallel push–pull operation and anode pulse modulation, powers in the MW range were attained.

Figure 21: The power triode TS6 was a typical development by Heinz Röhrig of GEMA. It was for a long time the standard transmitter output for the Seetakt sets working on 80 cm.

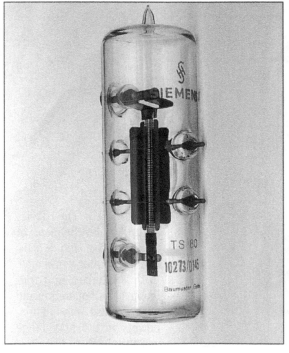

Figure 22: The TS60 delivered substantially higher power than the TS6 from which it was developed. Most of these tubes were manufactured by Siemens & Halske to the specifications furnished by GEMA.

178

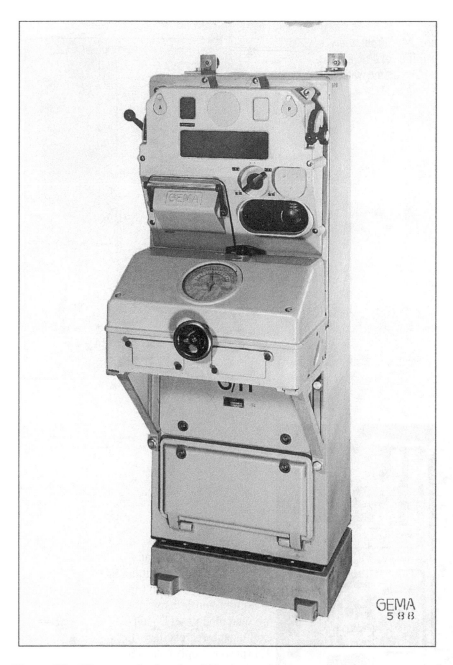

Figure 23: The console for the GH sonar with compensation was a typical example of GEMA's construction according to naval standards. From top to bottom: indicator, compensator, twin amplifier and power supply. In this sonar the transducers were mounted on the bow of a ship. Direction was determined from relative phase shift by the compensator.

Figure 24: Front view of the cabinet of Freya transmitter T with the watertight panels shut.

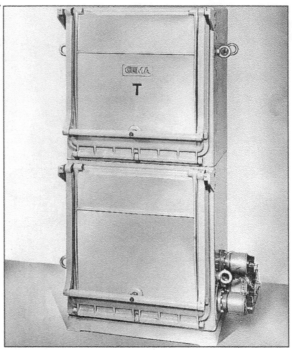

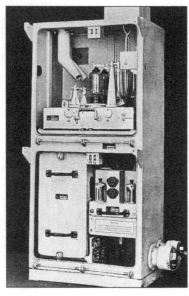

Figure 25: Front view of the cabinet of Freya transmitter T with the watertight panels open. From top to bottom and left to right are seen modules TU (self-exciting power oscillator), TN (power supply) and TS (control unit). The safety interlocks can be seen at the top of each unit.

Figure 26: DeTe-Panorama, the prototype of the radar that came to be called Jagdschloss. A 20 m long triangular frame held 18 full-wave dipoles and reflectors. The assembly rotated at 6 revolutions per minute. The maximum range was about 120 km. Presentation was by plan position indicator.

Figure 27: DeTe-Panorama seen mounted on a brick observation tower at Tremmen near Berlin.

Figure 28: A naval radar of the type Calais-B, on Cape d'Antifer on the French Channel coast. The antenna could be lowered hydraulically out of sight within the camouflage by the same mechanism as could the Limber Freya. British intelligence named this set Coast Watcher.

Figure 29: The GEMA Administration Building on Wendenschloss Strasse in Berlin-Köpenick.

Figure 30: The GEMA Berlin-Köpenick plant in a photograph taken some years after the war. Wendenschloss Strasse is on the right, the Dahme (a tributary of the River Spree) is on the left. The buildings in the foreground replaced the great assembly hall, which was demolished by the Soviet authorities. In the upper left is seen the experimental tower used for testing radar.

Figure 31: GEMA developed for the Luftwaffe this air-warning radar, Mammut. With its large antenna array of 30 m by 11 m it could detect aircraft at 300 km. It worked on the 1.5 m wavelength band and formed a very narrow vertical, fan-shaped radiation pattern. Phase compensators allowed the radiation pattern to be deflected ±50° from the central axis. British intelligence named it Hoarding (in American, billboard) after its appearance.

Figure 32: One of the tasks assigned this 60 m high GEMA Wassermann-S was the detection of British courier flights to Stockholm. It was an outstanding air-warning radar that attained enhanced range and accurate definition through the use of Freya electronics with a greatly increased antenna gain. This array of dipoles formed a thin, horizontal fan-shaped beam that could be moved up and down by altering the phase shifts in the transmission lines feeding the dipoles. This allowed a direct way of determining elevation to an accuracy of 0.75° for aircraft flying up to 25°; lobe switching gave an azimuthal accuracy of 0.25°. British intelligence named these sets Chimney because of its 'smokestack' appearance.

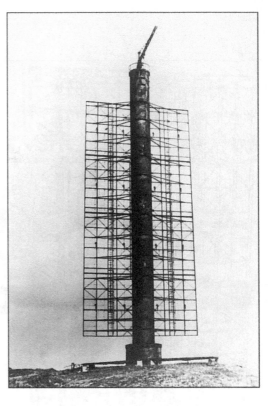

Figure 33: An experimental Limber Freya outfitted with a provisional antenna used for interrogating the IFF set Erstling. 1940.

186

Figure 34: This photograph shows the elaborate configuration of the antenna of a Wassermann-S. These sets were constructed at the site. A gear mechanism in a bunker allowed 360° rotation.

Figure 35: In 1941 radars were installed on U-boats that GEMA had developed earlier. The radiation pattern of the beam could be traversed by altering the phase shifts in the transmission lines feeding the dipoles. The lower array was for transmission, the upper for reception.

Figure 36: GEMA entered this antiaircraft gun-laying radar as a competitor for the Telefunken and Lorenz Flak radars, which it equalled or exceeded in accuracy. It employed the 80 cm Seetakt wavelength and could be used against both sea and air targets. It employed lobe switching for elevation and bearing. It was not adopted by the Luftwaffe but was employed by naval antiaircraft and was mounted on the cruiser *Prinz Eugen*.

Figure 37: One of the most successful of GEMA's radar designs was this Freya-LZ. In contrast to the Limber Freya it was of lighter and demountable construction, which allowed transport by air. It was assembled for operation after transport by whatever means available. It is shown here with three antennas: the IFF at the top, the receiver in the middle and the transmitter at the bottom. British intelligence named it the Pole Freya.

Figure 38: This photograph shows a Freya-LZ with broad-band dipoles for use with the Wismar modification that allowed small changes of frequency to evade jamming. Note that the polarization of the radar antennas is horizontal. This set was also used in a 360° panoramic mode as the Dreh-Freya.

Figure 39: The former cloister at Liegnitz in which GEMA set up some of its operations to escape the effects of air attacks on Berlin. It was known as the Blüchwerke.

Figure 40: A grain-storage hall at Lensahn where GEMA fled during the last weeks of the war when Liegnitz had been occupied by the Russians and the fall of Berlin was imminent.

Figure 41: A factory for sliced turnips at Lensahn which GEMA also used during the last weeks of the war.

Figure 42: Even leading GEMA officials had to pass through security. Here is a copy of the pass for Dr Walter Brandt, leader of the low-frequency lab at the Blücherwerke.

Figure 43: The author is shown at the former Institute for Ionosphere Research where he worked in 1952 in adapting GEMA transmitters using spark modulators for ionospheric sounding.

NAME INDEX

SUBJECT INDEX

Printed and bound by CPI Group (UK) Ltd, Croydon, CR0 4YY

23/10/2024

01778249-0007